Communications and Control Engineering

Springer
London
Berlin
Heidelberg
New York
Hong Kong
Milan
Paris
Tokyo

http://www.springer.de/engine/

Ole Morten Aamo and Miroslav Krstić

Flow Control by Feedback

Stabilization and Mixing

With 81 Figures

To Hassan,
With my best compliments,
Miroslav

 Springer

Ole Morten Aamo, PhD
Norwegian University of Science and Technology, N-7491 Trondheim, Norway

Miroslav Krstić, PhD
University of California, San Diego, La Jolla, CA 92093-0411, USA

Series Editors
E.D. Sontag • M. Thoma • A. Isidori • J.H. van Schuppen

British Library Cataloguing in Publication Data
Aamo, Ole Morten
    Flow control by feedback : stabilization and mixing. –
    (Communications and control engineering)
    1. Hydraulic control 2. Fluid mechanics
    I. Title II. Krstić, Miroslav
    620.1'064
ISBN 1852336692

Library of Congress Cataloging-in-Publication Data
A catalog record for this book is available from the Library of Congress

Apart from any fair dealing for the purposes of research or private study, or criticism or review, as permitted under the Copyright, Designs and Patents Act 1988, this publication may only be reproduced, stored or transmitted, in any form or by any means, with the prior permission in writing of the publishers, or in the case of reprographic reproduction in accordance with the terms of licences issued by the Copyright Licensing Agency. Enquiries concerning reproduction outside those terms should be sent to the publishers.

Communications and Control Engineering Series ISSN 0178-5354
ISBN 1-85233-669-2 Springer-Verlag London Berlin Heidelberg
A member of BertelsmannSpringer Science+Business Media GmbH
http://www.springer.co.uk

© Springer-Verlag London Limited 2003
Printed in Great Britain

The use of registered names, trademarks, etc. in this publication does not imply, even in the absence of a specific statement, that such names are exempt from the relevant laws and regulations and therefore free for general use.

The publisher makes no representation, express or implied, with regard to the accuracy of the information contained in this book and cannot accept any legal responsibility or liability for any errors or omissions that may be made.

Typesetting: Electronic text files prepared by author
Printed and bound at the Athenæum Press Ltd., Gateshead, Tyne & Wear
69/3830-543210 Printed on acid-free paper SPIN 10884668

Preface

In the 70+ year history of control theory and engineering, few applications have stirred as much excitement as flow control. The same can probably be said for the general area of fluid mechanics with its much longer history of several centuries. This excitement is understandable and justified. Turbulence in fluid flows has been recognized as the last great unsolved problem of classical physics[1] and has driven the careers of many leading mathematicians of the 20th century.[2] Likewise, control theorists have hardly ever come across a problem this challenging.

The emergence of flow control as an attractive new field is owed to the break-throughs in micro-electro-mechanical systems (MEMS) and other technologies for instrumenting fluid flows on extremely short length and time scales. The remaining missing ingredient for turning flow control into a practical tool is control algorithms with provable performance guarantees. This research monograph is the first book dedicated to this problem—systematic *feedback* design for fluid flows.

We are aware of the great interest in the topic among both control theorists, fluid mechanicists, and even mathematicians and physicists. This is why the book contains preliminaries to make its content accessible to graduate student level readers from all these fields. For the benefit of controls students we have included a self contained introduction on Navier-Stokes equations. To assist fluid dynamicists with control concepts, we have added an overview of some of the basics in linear and nonlinear control. To help non-engineers we have provided some elementary information on the current sensor/actuator/MEMS technology.

---

[1] A statement attributed to, among others, Einstein and Feynman, albeit undocumented [68].

[2] Well posedness of 3D Navier-Stokes partial differential equations with large initial data and high Reynolds number remains an open problem.

The scope of efforts in flow control is extremely broad. Control engineers are typically used to stabilization/regulation/tracking/disturbance attenuation problems where the quantity being minimized is well defined. In flow control one finds both such problems and the problems that are exactly opposite—where one has the task of destabilizing the system. Mixing of fluids is such a problem. Mixing is achieved by generating turbulence and is beneficial in several applications: fuel/air mixing in combustion, noise and infrared signature reduction at jet engine exhaust, and mixing of reactants in chemical process industry. This monograph dedicates a substantial portion to control of mixing and pioneers feedback concepts in this traditionally open-loop area. Both stabilization and destabilization, i.e., both relaminarization and turbulence enhancement, are treated in the book in a unified way.

While we give our own results in more detail than the results that precede them, the book is fairly complete in surveying the principal feedback algorithms for flow control available at present time.[3] An instructor can therefore use the book as a text in a stand-alone course on flow control, or as a supplemental text in courses on fluid dynamics or control of infinite dimensional systems.

The core of the book are Chapters 4 and 5. The lengthy Chapter 4 covers stabilization techniques, including linear optimal control results for discretized models, Lyapunov results for full Navier-Stokes models, and some backstepping results. Chapter 5 covers mixing control techniques, including approaches inspired by dynamical systems theory and Lyapunov-based inverse optimal approaches. The range of geometries covered includes 2D and 3D channel flow, pipe flow, and bluff body (cylinder) flow.

The material based on the first author's dissertation work are Section 4.3 (except Sections 4.3.2 and 4.3.3), Section 4.4, and Sections 5.2–5.4.

Acknowledgements

We owe great gratitude to our coauthors in works leading to this book: Andras Balogh, Weijiu Liu, and Thomas Bewley. In addition, we have benefited from support from or interaction with Kishan Baheti, John Baillieul, Bassam Bamieh, Andrzej Banaszuk, Peter Blossey, Jean-Michel Coron, Olav Egeland, Thor I. Fossen, George Haller, Marc Jacobs, Belinda King, Petar Kokotović, Juan Lasheras, Eric Lauga, Igor Mezić, Peter Monkewitz, Bartosz Protas, Gabriel Roy, Svein Ivar Sagatun, Yong Wang, Forman Williams, and Lawrence Yuan.

We gratefully acknowledge the support that we have received from the Air Force Office of Scientific Research, the National Science Foundation, the Norwegian Research Council, Norsk Hydro, and the Office of Naval Research.

---

[3] We specifically do not cover the results on controllability of Navier-Stokes equations (Fursikov, Imanuvilov, Coron, etc.). These are open-loop mathematical existence results of great significance but are not in a form implementable numerically or experimentally.

Finally, we thank our greatest supporters: Linda, Anna, Oline, Angela, Alexandra, and Victoria.

Trondheim, Norway   OLE MORTEN AAMO
La Jolla, California   MIROSLAV KRSTIĆ
May 2002

Contents

Chapter 1

Introduction

1.1 Why Flow Control?

Flow control involves controlling a flow field using passive or active devices in order to bring on desired changes in the behavior of the flow. For instance, *laminar* flow, which is characterized by parallel layers of fluid moving in a very regular and deterministic way, is associated with considerable less drag, or friction, at wall-fluid interfaces, than its counterpart, *turbulent* flow, which is characterized by small scale velocity components that appear to be stochastic in nature. On the other hand, turbulent flow may exhibit better mixing properties than laminar flows. Usually, laminar flows are unstable, and will unless controlled, evolve into turbulent flows. Common control objectives include [54]:

- Delaying or advancing transition from laminar to turbulent flow;
- Suppressing or enhancing turbulence, and;
- Preventing or provoking separation.

The benefits that can be gained from these control objectives include drag reduction, lift enhancement, mixing enhancement, and flow-induced noise suppression. For example, a turbulent pipe flow induces considerable drag, or friction, at the bounding wall. It is the resulting overall drag force that the compressor has to overcome in order to pump fluid through the pipe. Increasing the throughput can be achieved by simply installing a more powerful compressor, but the result would be increased energy consumption. Since laminar flow induces much less drag, designing a flow control system that relaminarizes the flow will permit higher throughput without increasing energy consumption.

In flows past bluff bodies, the phenomenon of vortex shedding occurs. For flow past a 2D circular cylinder, which is a prototype model flow for studying vortex shedding, vortices are alternatively shed from the upper and lower sides of the cylinder, subjecting the cylinder to periodic forcing. In practice the periodic forcing leads to structural vibrations, which are associated with penalties ranging from passenger discomfort to structural damage or failure from fatigue. Consequently, suppression of vortex shedding is of great importance in many engineering applications. By applying flow control, one may alter the behavior of the flow around the structure in such a manner that vortex shedding is suppressed or dampened.

Mixing processes are encountered frequently in applications, and the quality of the resulting mixture directly affects the quality of the end product. This is the case, for instance, in combustion, where the quality of the fuel-air mixture is essential for power generation, and in process industry, where the quality of various mixtures affect chemical reaction rates and the purity of end products. Mixing is usually obtained using "brute-force" techniques, such as mechanical stirring, jet injection and stirring valves. These methods, and all other methods for mixing, are associated with a drag penalty. The application of flow control to mixing problems seeks to minimize this penalty.

The feedback control laws designed for these problems are allowed to be distributed. That is, sensing and actuation may be applied at every point on the boundary of the flow domain. While this may sound unrealistic, the micromachining technology that emerged in the 1980s, permits rapid sensing and actuation on the micron scale, and thereby enables real-time distributed control of fluid flows.

It is clear from the examples mentioned above that the main objective in flow control is to lower operational expenses. We conclude this motivational section with a quote from [54]: "The potential benefits of realizing efficient flow-control systems range from saving billions of dollars in annual fuel costs for land, air, and sea vehicles to achieving economically and environmentally more competitive industrial processes involving fluid flows."

1.2   Scope of this Monograph

The concept of flow control contains a wide variety of theoretical and technological branches, as the previous section suggests. In this monograph the treatment is limited to recent developments in two problem areas that have attracted much attention. The first is *stabilization* of a selection of popular prototype flows, namely the channel flow (2D or 3D), the pipe flow, and the cylinder flow (2D). The channel flow is the flow contained between two parallel plates, the pipe flow is the flow contained in the interior of a cylinder with circular cross section, and the cylinder flow is the flow past a cylinder with circular cross section. The second problem we will study, is *mixing* in flows. In the next two sections, recent efforts in these areas are reviewed briefly. Then, selected works are treated in detail in Chapters 4 and 5. First, though, the equations of fluid mechanics are reviewed in Chapter 2, and some control theoretic results are reviewed in Chapter 3. Following the core chapters on stabilization and mixing, a review of the state-of-the-art in sensing and actuation for fluid mechanical systems is presented in Chapter 6.

1.2.1 Stabilization

Incompressible fluid flow in a plane channel has been studied quite extensively, and the wall sensing/actuation of this flow has become a standard benchmark problem in the area of flow control, see, e.g., [25] and [53] for recent reviews. In [78] stabilizing PI controllers for two-dimensional channel flow were designed for a reduced-order model of the linearized Navier-Stokes equation, obtained by a standard Galerkin procedure. The work was continued in [79], where LQG design was applied in order to obtain optimal controllers for this reduced-order model. LQG/LTR of the streamfunction formulation of the Navier-Stokes equation was also the focus of [43] and [42], where the latter reference reports the remarkable result of drag reduction to 50% below the laminar level. In [83], a reduced-order model of 3D perturbations (at a single wavenumber pair $\{k_x, k_z\}$) was developed, and LQG control design was applied to this model. Drag reduction by means of body forcing inside the domain applied through electromagnetic forcing was suggested in [14], where an observer-based approach was applied to a reduced order, linearized model. A nonlinear attempt was presented in [38], where Galerkin's method was used to derive a reduced-order model of the full, nonlinear, two-dimensional Navier-Stokes equation. A nonlinear control law was given, along with conditions under which closed-loop stability is obtained. The results were applied to Burger's equation, which is the one-dimensional Navier-Stokes equation, including the nonlinear advective term. In [24], LQG and $\mathcal{H}_\infty$ control theories were applied to the linearized three-dimensional channel flow. A major finding of this paper was that properly-applied controls significantly reduce the nonorthogonality leading to energy amplification mechanisms in such systems. The three-dimensional nonlinear problem was tackled by the application of optimal control theory in a finite-horizon predictive setting (Model Predictive Control) in [27], resulting in relaminarization of $Re = 1700$ turbulent channel flow. In [21], the authors considered the externally excited linearized Navier-Stokes equation, and employed the tools of robust control for resolving some long standing open problems in fluid dynamics related to energy amplification in flows. Using the same input-output setting for analysis of the externally excited linearized Navier-Stokes equation, an accurate statistical model of real turbulent flow was developed in [80, 81]. It was proposed that the use of this model in an LQG control problem for channel flow will result in superior performance compared to designs based on spatially and temporally white noise as external excitation.

Optimal controllers are generally not decentralized, but recent results on the structure of controllers for spatially-invariant systems indicate that one can obtain *localized* controllers arbitrarily close to optimal [19]. This result has been confirmed for plane channel flow in [72], where the Fourier-space control problem formulated in [24] was modified and successfully inverse-transformed to the physical domain, resulting in well-resolved, spatially-localized convolution kernels with exponential decay far from the origin. Such spatial localization is an important ingredient both in relaxing the nonphysical assumption of spatial periodicity in the controller formulation and in facilitating decentralized control in massive arrays of sensors and actuators (see discussion in [25]). The mathematical details of controllability and optimal control theory applied to the Navier-Stokes equation, such as existence and uniqueness of solutions, and proofs of convergence of proposed numerical algorithms, are discussed in [7, 22, 23, 28, 40, 41, 47, 48, 50, 58, 59, 69, 73, 74, 94, 124, 125].

Global stabilization by boundary control of Burgers' equation was achieved in [91]. For the 2D channel flow governed by the Navier-Stokes equation, globally stabilizing boundary control laws were presented in [16], where wall-tangential actuation was used, and in [1], where wall-normal actuation was used. These control laws were fully decentralized, and numerical simulations showed their ability to stabilize flows at large Reynolds numbers, although the mathematical analysis was valid for small Reynolds numbers, only. As noted in [16] and [25], fully decentralized controllers have an implementational advantage in that they can be embedded into MEMS (Micro-Electro-Mechanical-Systems) hardware, minimizing the communication requirements of centralized computations and facilitating scaling to massive arrays of sensors and actuators. Using the wall-tangential control law, Balogh [18] was able to relaminarize a simulated turbulent 3D channel flow at $Re = 4000$.

The flow past a 2D circular cylinder has been a popular model flow for studying vortex suppression by means of open-loop or feedback control. For Reynolds numbers slightly larger than the critical value for onset of vortex shedding (which is approximately $Re_c = 47$), several authors have successfully suppressed vortex shedding in numerical simulations using various simple feedback control configurations. In [113], a pair of suction/blowing slots positioned on the cylinder wall were used for actuation, and shedding was suppressed for $Re = 60$, using proportional feedback from a single velocity measurement taken some distance downstream of the cylinder. For $Re = 80$, vortex shedding was reduced, but not completely suppressed. In [60], the same actuation configuration was tried using feedback from a pair of pressure sensors located on the cylinder wall for $Re = 60$. This attempt was unsuccessful, but by adding a third actuation slot, shedding was reduced considerably, even at $Re = 80$. Although some success in controlling vortex shedding has been achieved in numerical simulations, rigorous control designs are scarce due to the complexity of designing controllers based on the Navier-Stokes equation. A much simpler model, the Ginzburg-Landau equation with appropriate coefficients, has been found to model well the dynamics of vortex shedding near the critical value of the Reynolds number

[71]. In [117], it was shown numerically that the Ginzburg-Landau model for Reynolds numbers close to Re_c can be stabilized using proportional feedback from a single measurement downstream of the cylinder, to local forcing at the location of the cylinder. In [95], using the model from [117], stabilization was obtained in numerical simulations for $Re = 100$, with an LQG controller designed for the linearized Ginzburg-Landau equation. In [4], a controller that globally stabilized the equilibrium at zero of a finite difference discretization of any order of the nonlinear Ginzburg-Landau model presented in [117] was designed using backstepping. The design was valid for any Reynolds number.

1.2.2 Mixing

In many engineering applications, the mixing of two or more fluids is essential to obtaining good performance in some downstream process (a prime example is the mixing of air and fuel in combustion engines [55, 11]). As a consequence, mixing has been the focus of much research, but without reaching a unified theory, either for the generation of flows that mix well due to external forcing, or for the quantification of mixing in such flows. A thorough review is given in [110]. Approaches range from experimental design and testing to modern applications of dynamical systems theory. The latter was initiated by Aref [12], who studied chaotic advection in the setting of an incompressible, inviscid fluid contained in a (2D) circular domain, and agitated by a point vortex. This flow is commonly called the blinking vortex flow. Ottino and coworkers studied a number of various flows, examining mixing properties based on dynamical systems techniques [37, 85, 97, 126]. Later Rom-Kedar et al. [116] applied Melnikov's method and KAM (Kolmogorov-Arnold-Moser) theory to quantify transport in a flow governed by an oscillating vortex pair. For general treatments of dynamical systems theory and transport in dynamical systems, see [57, 133, 134]. An obvious shortcoming of this theory is the requirement that the flow must be periodic, as such methods rely on the existence of a Poincaré map for which some periodic orbit of the flow induces a hyperbolic fixed point. Another shortcoming is that they can only handle small perturbations from integrability, whereas effective mixing usually occurs for large perturbations [111]. A third shortcoming is that traditional dynamical systems theory is concerned with asymptotic, or long-time, behavior, rather than quantifying rate processes which are of interest in mixing applications. In order to overcome some of these shortcomings, recent advances in dynamical systems theory have focused on finding coherent structures and invariant manifolds in experimental datasets, which are finite in time and generally aperiodic. This has led to the notions of finite-time hyperbolic trajectories with corresponding finite-time stable and unstable manifolds [62, 63, 64, 65]. The results include estimates for the transport of initial conditions across the boundaries of coherent structures. In [115] these concepts were applied to a time-dependent velocity field generated by a double-gyre ocean model, in order to study the fluid transport between dynamic eddies and a jet stream. An application to meandering jets was described in [107]. Another method for identifying regions in a flow that have similar finite-time statistical properties based on ergodic theory was developed and applied in [104, 102, 105]. The relationship between the two methods mentioned, focusing on geometrical and statistical properties of particle motion, respectively, was examined in [114].

As these developments have partly been motivated by applications in geophysical flows, they are diagnostic in nature and lend little help to the problem of *generating* a fluid flow that mixes well. The problem of generating effective mixing in a fluid flow is usually approached by trial and error using various "brute force" open-loop controls, such as mechanical stirring, jet injection or mixing valves. However, in the recent papers [44, 45], control systems theory was used to rigorously derive the mixing protocol that maximizes entropy among all the possible periodic sequences composed of two shear flows orthogonal to each other. In [108], the optimal vortex trajectory in the flow induced by a single vortex in a corner subject to a controlled external strain field was found using tools from dynamical systems theory. The resulting trajectory was stabilized using control theory.

In [1], feedback control was applied in order to enhance existing instability mechanisms in a 2D model of plane channel flow. By applying boundary control intelligently in a feedback loop, mixing was considerably enhanced with relatively small control effort. The control law was decentralized and designed using Lyapunov stability analysis. These efforts have recently been extended successfully to 3D pipe flow [17], where certain optimality properties were shown as well. In the recent paper [20], the authors present a framework for destabilization, in an optimal manner, of linear time-invariant systems for the purpose of achieving mixing enhancement. Motivated by the results in [1, 17], a simulation study was carried out in [6], aiming at enhancing particle dispersion in the wake of a 2D cylinder. For the subcritical case of $Re = 45$, vortex shedding was successfully initiated using feedback control.

Chapter 2

Governing Equations

2.1 Kinematics

We will be studying the behavior of a fluid contained in the domain Ω, as shown schematically in Figure 2.1. Associated with the fluid is its density, $\rho : \Omega \times \mathbb{R}_+ \rightarrow \mathbb{R}$. At every time instant $t > 0$, and to every point $p \in \Omega$, we assign a vector valued quantity which is the velocity, $\mathbf{W}$, of the fluid at that point in time and space. That is, we are interested in the evolution of a vector field $\mathbf{W} : \Omega \times \mathbb{R}_+ \rightarrow \mathbb{R}^n$, where n is the *dimension* of the problem. Associated with the velocity field is a pressure field, which is a scalar valued function $P : \Omega \times \mathbb{R}_+ \rightarrow \mathbb{R}$. We will study problems in 2 and 3 dimensions (2D and 3D), using cartesian and cylindrical coordinates. In cartesian coordinates, we denote a point $p \in \Omega$ with (x, y) in 2D and (x, y, z) in 3D. In cylindrical coordinates, we denote a point $p \in \Omega$ with (r, θ, z). The two coordinate systems are shown schematically in Figure 2.1. The velocity field is denoted $\mathbf{W}(x, y, z, t) = (U(x, y, z, t), V(x, y, z, t), W(x, y, z, t))$ in 3D cartesian coordinates, where U, V and W are the velocity components in the x, y and z directions, respectively ($\mathbf{W}(x, y, t) = (U(x, y, t), V(x, y, t))$ in 2D). In cylindrical coordinates, we denote the velocity field $\mathbf{W}(r, \theta, z, t) = (V_r(x, y, z, t), V_\theta(x, y, z, t), V_z(x, y, z, t))$, where V_r, V_θ, and V_z are the velocity components in the r, θ and z directions, respectively. The density, ρ, and the pressure, P, take the same arguments as the velocity, but are scalar valued. Below we will derive the conservation equations in cartesian coordinates, and state the corresponding equations in cylindrical coordinates. The derivation follows [31].

2.2 Conservation of Mass

Consider the stationary volume element in Figure 2.2. Writing a mass balance

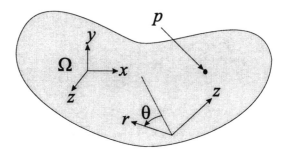

Figure 2.1: The domain in which the fluid is contained is denoted Ω. Two coordinate systems will be used in this report: cartesian coordinates, denoted (x, y, z), and cylindrical coordinates, denoted (r, θ, z).

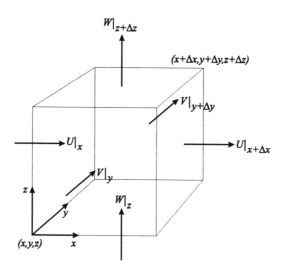

Figure 2.2: Control volume for derivation of the governing equations.

over the volume, we have

rate of mass accumulation = rate of mass in − rate of mass out

where

$$\text{rate of mass in} = (\rho U)|_x \Delta y \Delta z + (\rho V)|_y \Delta x \Delta z + (\rho W)|_z \Delta x \Delta y$$
$$\text{rate of mass out} = (\rho U)|_{x+\Delta x} \Delta y \Delta z + (\rho V)|_{y+\Delta y} \Delta x \Delta z$$
$$+ (\rho W)|_{z+\Delta z} \Delta y \Delta z.$$

Thus, we get

$$\frac{\partial \rho}{\partial t} \Delta x \Delta y \Delta z = - \left((\rho U)|_{x+\Delta x} - (\rho U)|_x \right) \Delta y \Delta z$$
$$- \left((\rho V)|_{y+\Delta y} - (\rho V)|_y \right) \Delta x \Delta z - \left((\rho W)|_{z+\Delta z} - (\rho W)|_z \right) \Delta y \Delta z.$$

Dividing by the volume and letting Δx, Δy, and Δz approach zero, we get

$$\frac{\partial \rho}{\partial t} = -\frac{\partial (\rho U)}{\partial x} - \frac{\partial (\rho V)}{\partial y} - \frac{\partial (\rho W)}{\partial z}.$$

We will be dealing exclusively with incompressible fluids, for which ρ is constant. Consequently,

$$\frac{\partial U}{\partial x} + \frac{\partial V}{\partial y} + \frac{\partial W}{\partial z} = 0. \tag{2.1}$$

Equation (2.1) is referred to as the *equation of continuity*.

2.3 Conservation of Momentum

Again consider the stationary volume element in Figure 2.2. Writing a momentum balance over the volume, we have

rate of momentum accumulation =
rate of momentum in − rate of momentum out
+ sum of forces acting on system.

So for the momentum in the x-direction, we have

rate of momentum in =

$$(\rho U^2)|_x \Delta y \Delta z + (\rho U V)|_y \Delta x \Delta z + (\rho U W)|_z \Delta x \Delta y$$

rate of momentum out =

$$(\rho U^2)|_{x+\Delta x} \Delta y \Delta z + (\rho U V)|_{y+\Delta y} \Delta x \Delta z + (\rho U W)|_{z+\Delta z} \Delta x \Delta y$$

sum of forces acting on system $=$

$$\left(\tau_{xx}|_x - \tau_{xx}|_{x+\Delta x}\right)\Delta y\Delta z + \left(\tau_{yx}|_y - \tau_{yx}|_{y+\Delta y}\right)\Delta x\Delta z$$
$$+ \left(\tau_{zx}|_z \Delta x\Delta y - \tau_{zx}|_{z+\Delta z}\right)\Delta x\Delta y + \left(P|_x - P|_{x+\Delta x}\right)\Delta y\Delta z$$

where τ_{ij} denotes the viscous force (per unit area) acting in the direction of j on a face normal to the i-direction, and P is the pressure (i.e. the pressure force per unit area). We get

$$\frac{\partial(\rho U)}{\partial t}\Delta x\Delta y\Delta z = -\left(\left(\rho U^2\right)|_{x+\Delta x} - \left(\rho U^2\right)|_x\right)\Delta y\Delta z$$
$$-\left(\left(\rho UV\right)|_{y+\Delta y} - \left(\rho UV\right)|_y\right)\Delta x\Delta z - \left(\left(\rho UW\right)|_{z+\Delta z} - \left(\rho UW\right)|_z\right)\Delta x\Delta y$$
$$-\left(\tau_{xx}|_{x+\Delta x} - \tau_{xx}|_x\right)\Delta y\Delta z - \left(\tau_{yx}|_{y+\Delta y} - \tau_{yx}|_y\right)\Delta x\Delta z$$
$$-\left(\tau_{zx}|_{z+\Delta z} - \tau_{zx}|_z\right)\Delta x\Delta y - \left(P|_{x+\Delta x} - P|_x\right)\Delta y\Delta z.$$

Dividing by the volume and letting Δx, Δy, and Δz approach zero, we get

$$\frac{\partial(\rho U)}{\partial t} + \frac{\partial(\rho U^2)}{\partial x} + \frac{\partial(\rho UV)}{\partial y} + \frac{\partial(\rho UW)}{\partial z} = -\frac{\partial P}{\partial x} - \frac{\partial\tau_{xx}}{\partial x} - \frac{\partial\tau_{yx}}{\partial y} - \frac{\partial\tau_{zx}}{\partial z}.$$

For constant density ρ, we get

$$\rho\frac{\partial U}{\partial t} + \rho\left(U\frac{\partial U}{\partial x} + V\frac{\partial U}{\partial y} + W\frac{\partial U}{\partial z}\right) + \rho U\left(\frac{\partial U}{\partial x} + \frac{\partial V}{\partial y} + \frac{\partial W}{\partial z}\right) =$$
$$-\frac{\partial P}{\partial x} - \frac{\partial\tau_{xx}}{\partial x} - \frac{\partial\tau_{yx}}{\partial y} - \frac{\partial\tau_{zx}}{\partial z}$$

and using (2.1) yields

$$\rho\frac{\partial U}{\partial t} + \rho\left(U\frac{\partial U}{\partial x} + V\frac{\partial U}{\partial y} + W\frac{\partial U}{\partial z}\right) = -\frac{\partial P}{\partial x} - \frac{\partial\tau_{xx}}{\partial x} - \frac{\partial\tau_{yx}}{\partial y} - \frac{\partial\tau_{zx}}{\partial z}.$$

Similar derivations for the momentum in the y and z directions yield the complete set of equations

$$\rho\frac{\partial U}{\partial t} + \rho\left(U\frac{\partial U}{\partial x} + V\frac{\partial U}{\partial y} + W\frac{\partial U}{\partial z}\right) = -\frac{\partial P}{\partial x} - \frac{\partial\tau_{xx}}{\partial x} - \frac{\partial\tau_{yx}}{\partial y} - \frac{\partial\tau_{zx}}{\partial z} \quad (2.2)$$

$$\rho\frac{\partial V}{\partial t} + \rho\left(U\frac{\partial V}{\partial x} + V\frac{\partial V}{\partial y} + W\frac{\partial V}{\partial z}\right) = -\frac{\partial P}{\partial y} - \frac{\partial\tau_{xy}}{\partial x} - \frac{\partial\tau_{yy}}{\partial y} - \frac{\partial\tau_{zy}}{\partial z} \quad (2.3)$$

$$\rho\frac{\partial W}{\partial t} + \rho\left(U\frac{\partial W}{\partial x} + V\frac{\partial W}{\partial y} + W\frac{\partial W}{\partial z}\right) = -\frac{\partial P}{\partial z} - \frac{\partial\tau_{xz}}{\partial x} - \frac{\partial\tau_{yz}}{\partial y} - \frac{\partial\tau_{zz}}{\partial z}. \quad (2.4)$$

It remains to insert a constitutive equation relating the viscous forces τ_{ij} to the fluid motion $\mathbf{W} = (U, V, W)$. We will exclusively consider Newtonian fluids, which in conjunction with incompressibility yield the following relations [31]

$$\tau_{xx} = -2\mu\frac{\partial U}{\partial x}, \; \tau_{yy} = -2\mu\frac{\partial V}{\partial y}, \; \tau_{zz} = -2\mu\frac{\partial W}{\partial z} \tag{2.5}$$

$$\tau_{xy} = \tau_{yx} = -\mu\left(\frac{\partial U}{\partial y} + \frac{\partial V}{\partial x}\right) \tag{2.6}$$

$$\tau_{yz} = \tau_{zy} = -\mu\left(\frac{\partial V}{\partial z} + \frac{\partial W}{\partial y}\right) \tag{2.7}$$

$$\tau_{zx} = \tau_{xz} = -\mu\left(\frac{\partial W}{\partial x} + \frac{\partial U}{\partial z}\right) \tag{2.8}$$

where the coefficient μ is called the viscosity of the fluid. Inserting (2.5)–(2.8) into (2.2)–(2.4) yields

$$\rho\frac{\partial U}{\partial t} + \rho\left(U\frac{\partial U}{\partial x} + V\frac{\partial U}{\partial y} + W\frac{\partial U}{\partial z}\right) =$$
$$-\frac{\partial P}{\partial x} + \mu\left(\frac{\partial^2 U}{\partial x^2} + \frac{\partial^2 U}{\partial y^2} + \frac{\partial^2 U}{\partial z^2}\right) + \mu\frac{\partial}{\partial x}\left(\frac{\partial U}{\partial x} + \frac{\partial V}{\partial y} + \frac{\partial W}{\partial z}\right)$$

$$\rho\frac{\partial V}{\partial t} + \rho\left(U\frac{\partial V}{\partial x} + V\frac{\partial V}{\partial y} + W\frac{\partial V}{\partial z}\right) =$$
$$-\frac{\partial P}{\partial y} + \mu\left(\frac{\partial^2 V}{\partial x^2} + \frac{\partial^2 V}{\partial y^2} + \frac{\partial^2 V}{\partial z^2}\right) + \mu\frac{\partial}{\partial y}\left(\frac{\partial U}{\partial x} + \frac{\partial V}{\partial y} + \frac{\partial W}{\partial z}\right)$$

$$\rho\frac{\partial W}{\partial t} + \rho\left(U\frac{\partial W}{\partial x} + V\frac{\partial W}{\partial y} + W\frac{\partial W}{\partial z}\right) =$$
$$-\frac{\partial P}{\partial z} + \mu\left(\frac{\partial^2 W}{\partial x^2} + \frac{\partial^2 W}{\partial y^2} + \frac{\partial^2 W}{\partial z^2}\right) + \mu\frac{\partial}{\partial z}\left(\frac{\partial U}{\partial x} + \frac{\partial V}{\partial y} + \frac{\partial W}{\partial z}\right)$$

and using (2.1) we finally get

$$\rho\left(\frac{\partial U}{\partial t} + U\frac{\partial U}{\partial x} + V\frac{\partial U}{\partial y} + W\frac{\partial U}{\partial z}\right) =$$
$$-\frac{\partial P}{\partial x} + \mu\left(\frac{\partial^2 U}{\partial x^2} + \frac{\partial^2 U}{\partial y^2} + \frac{\partial^2 U}{\partial z^2}\right) \tag{2.9}$$

$$\rho\left(\frac{\partial V}{\partial t} + U\frac{\partial V}{\partial x} + V\frac{\partial V}{\partial y} + W\frac{\partial V}{\partial z}\right) =$$
$$-\frac{\partial p}{\partial y} + \mu\left(\frac{\partial^2 V}{\partial x^2} + \frac{\partial^2 V}{\partial y^2} + \frac{\partial^2 V}{\partial z^2}\right) \tag{2.10}$$

$$\rho\left(\frac{\partial W}{\partial t} + U\frac{\partial W}{\partial x} + V\frac{\partial W}{\partial y} + W\frac{\partial W}{\partial z}\right) =$$

$$-\frac{\partial P}{\partial z} + \mu\left(\frac{\partial^2 W}{\partial x^2} + \frac{\partial^2 W}{\partial y^2} + \frac{\partial^2 W}{\partial z^2}\right). \quad (2.11)$$

Equation (2.9)–(2.11) is the celebrated *Navier-Stokes equation*. In vector form, equations (2.1) and (2.9)–(2.11) can be written compactly as

$$\boxed{div\,(\mathbf{W}) = 0}$$

$$\boxed{\frac{\partial \mathbf{W}}{\partial t} + (\mathbf{W} \cdot \nabla)\,\mathbf{W} = -\frac{1}{\rho}\nabla P + \frac{\mu}{\rho}\Delta \mathbf{W}}$$

where ∇ denotes the gradient operator, Δ denotes the Laplace operator, and *div* is short for divergence.

2.4   The Dimensionless Navier-Stokes Equation

2.4.1   Cartesian Coordinates

Given a flow geometry, the Navier-Stokes equation can be written in dimensionless form by introducing a characteristic length and a characteristic velocity. Denoting the characteristic length $\check{D}$, and the characteristic velocity $\check{V}$, we can perform a change of variables in such a way that the new variables are dimensionless

$$(U^*, V^*, W^*) = \left(\frac{U}{\check{V}}, \frac{V}{\check{V}}, \frac{W}{\check{V}}\right), P^* = \frac{P}{\rho\check{V}^2}, t^* = \frac{\check{V}}{\check{D}}t$$

$$x^* = \frac{x}{\check{D}}, y^* = \frac{y}{\check{D}}, z^* = \frac{z}{\check{D}}.$$

So we get

$$\frac{\partial U}{\partial t} = \check{V}\frac{\partial U^*}{\partial t^*}\frac{\partial t^*}{\partial t} = \frac{\check{V}^2}{\check{D}}\frac{\partial U^*}{\partial t^*}$$

$$\frac{\partial U}{\partial x} = \check{V}\frac{\partial U^*}{\partial x^*}\frac{\partial x^*}{\partial x} = \frac{\check{V}}{\check{D}}\frac{\partial U^*}{\partial x^*}$$

$$\frac{\partial P}{\partial x} = \rho\check{V}^2\frac{\partial P^*}{\partial x^*}\frac{\partial x^*}{\partial x} = \frac{\rho\check{V}^2}{\check{D}}\frac{\partial P^*}{\partial x^*}$$

$$\frac{\partial^2 U}{\partial x^2} = \frac{\partial}{\partial x^*}\left(\frac{\check{V}}{\check{D}}\frac{\partial U^*}{\partial x^*}\right)\frac{\partial x^*}{\partial x} = \frac{\check{V}}{\check{D}^2}\frac{\partial^2 U^*}{\partial x^{*2}}$$

and similarly for the other derivatives occurring in (2.1) and (2.9)–(2.11). Inserting these relations into (2.1) and (2.9)–(2.11) yields

$$\frac{\partial U}{\partial x} + \frac{\partial V}{\partial y} + \frac{\partial W}{\partial z} = 0 \tag{2.12}$$

$$\frac{\partial U}{\partial t} + U\frac{\partial U}{\partial x} + V\frac{\partial U}{\partial y} + W\frac{\partial U}{\partial z} =$$
$$-\frac{\partial P}{\partial x} + \frac{1}{Re}\left(\frac{\partial^2 U}{\partial x^2} + \frac{\partial^2 U}{\partial y^2} + \frac{\partial^2 U}{\partial z^2}\right) \tag{2.13}$$

$$\frac{\partial V}{\partial t} + U\frac{\partial V}{\partial x} + V\frac{\partial V}{\partial y} + W\frac{\partial V}{\partial z} =$$
$$-\frac{\partial P}{\partial y} + \frac{1}{Re}\left(\frac{\partial^2 V}{\partial x^2} + \frac{\partial^2 V}{\partial y^2} + \frac{\partial^2 V}{\partial z^2}\right) \tag{2.14}$$

$$\frac{\partial W}{\partial t} + U\frac{\partial W}{\partial x} + V\frac{\partial W}{\partial y} + W\frac{\partial W}{\partial z} =$$
$$-\frac{\partial P}{\partial z} + \frac{1}{Re}\left(\frac{\partial^2 W}{\partial x^2} + \frac{\partial^2 W}{\partial y^2} + \frac{\partial^2 W}{\partial z^2}\right) \tag{2.15}$$

where we have defined $Re = \rho \check{D}\check{V}/\mu$, and skipped the superscript * for notational convenience. Re is called the Reynolds number, and is the only parameter in equations (2.13)–(2.15). In vector notation, equations (2.12) and (2.13)–(2.15) become

$$\boxed{div\,(\mathbf{W}) = 0}$$

$$\boxed{\frac{\partial \mathbf{W}}{\partial t} + (\mathbf{W} \cdot \nabla)\,\mathbf{W} = -\nabla P + \frac{1}{Re}\Delta\mathbf{W}}$$

2.4.2 Cylindrical Coordinates

In cylindrical coordinates, equations (2.12)–(2.15) become (see, for instance, [31]),

$$\frac{1}{r}\frac{\partial}{\partial r}\left(rV_r\right) + \frac{1}{r}\frac{\partial V_\theta}{\partial \theta} + \frac{\partial V_z}{\partial z} = 0 \tag{2.16}$$

$$\frac{\partial V_r}{\partial t} + V_r\frac{\partial V_r}{\partial r} + \frac{V_\theta}{r}\frac{\partial V_r}{\partial \theta} - \frac{V_\theta^2}{r} + V_z\frac{\partial V_r}{\partial z} =$$
$$-\frac{\partial P}{\partial r} + \frac{1}{Re}\left(\frac{\partial}{\partial r}\left(\frac{1}{r}\frac{\partial}{\partial r}\left(rV_r\right)\right) + \frac{1}{r^2}\frac{\partial^2 V_r}{\partial \theta^2} - \frac{2}{r^2}\frac{\partial V_\theta}{\partial \theta} + \frac{\partial^2 V_r}{\partial z^2}\right) \tag{2.17}$$

$$\frac{\partial V_\theta}{\partial t} + V_r \frac{\partial V_\theta}{\partial r} + \frac{V_\theta}{r} \frac{\partial V_\theta}{\partial \theta} + \frac{V_r V_\theta}{r} + V_z \frac{\partial V_\theta}{\partial z} =$$

$$-\frac{1}{r} \frac{\partial P}{\partial \theta} + \frac{1}{Re} \left(\frac{\partial}{\partial r} \left(\frac{1}{r} \frac{\partial}{\partial r} (r V_\theta) \right) + \frac{1}{r^2} \frac{\partial^2 V_\theta}{\partial \theta^2} + \frac{2}{r^2} \frac{\partial V_r}{\partial \theta} + \frac{\partial^2 V_\theta}{\partial z^2} \right) \quad (2.18)$$

$$\frac{\partial V_z}{\partial t} + V_r \frac{\partial V_z}{\partial r} + \frac{V_\theta}{r} \frac{\partial V_z}{\partial \theta} + V_z \frac{\partial V_z}{\partial z} =$$

$$-\frac{\partial P}{\partial z} + \frac{1}{Re} \left(\frac{1}{r} \frac{\partial}{\partial r} \left(r \frac{\partial V_z}{\partial r} \right) + \frac{1}{r^2} \frac{\partial^2 V_z}{\partial \theta^2} + \frac{\partial^2 V_z}{\partial z^2} \right). \quad (2.19)$$

2.5 Perturbations and the Linearized Navier-Stokes Equation

2.5.1 Cartesian Coordinates

Suppose $(\tilde{U}, \tilde{V}, \tilde{W}, \tilde{P})$ is a steady state solution of (2.12) and (2.13)–(2.15), that is

$$\left(\frac{\partial \tilde{U}}{\partial t}, \frac{\partial \tilde{V}}{\partial t}, \frac{\partial \tilde{W}}{\partial t}, \frac{\partial \tilde{P}}{\partial t} \right) = (0, 0, 0, 0) \quad (2.20)$$

and

$$\frac{\partial \tilde{U}}{\partial x} + \frac{\partial \tilde{V}}{\partial y} + \frac{\partial \tilde{W}}{\partial z} = 0 \quad (2.21)$$

$$\tilde{U} \frac{\partial \tilde{U}}{\partial x} + \tilde{V} \frac{\partial \tilde{U}}{\partial y} + \tilde{W} \frac{\partial \tilde{U}}{\partial z} = -\frac{\partial \tilde{P}}{\partial x} + \frac{1}{Re} \left(\frac{\partial^2 \tilde{U}}{\partial x^2} + \frac{\partial^2 \tilde{U}}{\partial y^2} + \frac{\partial^2 \tilde{U}}{\partial z^2} \right) \quad (2.22)$$

$$\tilde{U} \frac{\partial \tilde{V}}{\partial x} + \tilde{V} \frac{\partial \tilde{V}}{\partial y} + \tilde{W} \frac{\partial \tilde{V}}{\partial z} = -\frac{\partial \tilde{P}}{\partial y} + \frac{1}{Re} \left(\frac{\partial^2 \tilde{V}}{\partial x^2} + \frac{\partial^2 \tilde{V}}{\partial y^2} + \frac{\partial^2 \tilde{V}}{\partial z^2} \right) \quad (2.23)$$

$$\tilde{U} \frac{\partial \tilde{W}}{\partial x} + \tilde{V} \frac{\partial \tilde{W}}{\partial y} + \tilde{W} \frac{\partial \tilde{W}}{\partial z} = -\frac{\partial \tilde{P}}{\partial z} + \frac{1}{Re} \left(\frac{\partial^2 \tilde{W}}{\partial x^2} + \frac{\partial^2 \tilde{W}}{\partial y^2} + \frac{\partial^2 \tilde{W}}{\partial z^2} \right). \quad (2.24)$$

Defining the perturbation variables $\mathbf{w} = (u, v, w)$ and p as

$$u \triangleq U - \tilde{U}$$
$$v \triangleq V - \tilde{V}$$
$$w \triangleq W - \tilde{W}$$
$$p \triangleq P - \tilde{P}$$

and inserting into (2.12) and (2.13)–(2.15), yield

$$\frac{\partial u}{\partial x} + \frac{\partial v}{\partial y} + \frac{\partial w}{\partial z} + \frac{\partial \tilde{U}}{\partial x} + \frac{\partial \tilde{V}}{\partial y} + \frac{\partial \tilde{W}}{\partial z} = 0$$

$$\frac{\partial u}{\partial t} + \frac{\partial \tilde{U}}{\partial t} + u\frac{\partial u}{\partial x} + u\frac{\partial \tilde{U}}{\partial x} + \tilde{U}\frac{\partial u}{\partial x} + v\frac{\partial u}{\partial y} + v\frac{\partial \tilde{U}}{\partial y} + \tilde{V}\frac{\partial u}{\partial y}$$

$$+ w\frac{\partial u}{\partial z} + w\frac{\partial \tilde{U}}{\partial z} + \tilde{W}\frac{\partial u}{\partial z} + \tilde{U}\frac{\partial \tilde{U}}{\partial x} + \tilde{V}\frac{\partial \tilde{U}}{\partial y} + \tilde{W}\frac{\partial \tilde{U}}{\partial z} =$$

$$-\frac{\partial p}{\partial x} - \frac{\partial \tilde{P}}{\partial x} + \frac{1}{Re}\left(\frac{\partial^2 u}{\partial x^2} + \frac{\partial^2 u}{\partial y^2} + \frac{\partial^2 u}{\partial z^2} + \frac{\partial^2 \tilde{U}}{\partial x^2} + \frac{\partial^2 \tilde{U}}{\partial y^2} + \frac{\partial^2 \tilde{U}}{\partial z^2} \right)$$

$$\frac{\partial v}{\partial t} + \frac{\partial \tilde{V}}{\partial t} + u\frac{\partial v}{\partial x} + \tilde{U}\frac{\partial v}{\partial x} + \frac{\partial \tilde{V}}{\partial x}u + v\frac{\partial v}{\partial y} + v\frac{\partial \tilde{V}}{\partial y} + \tilde{V}\frac{\partial v}{\partial y}$$

$$+ w\frac{\partial v}{\partial z} + w\frac{\partial \tilde{V}}{\partial z} + \tilde{W}\frac{\partial v}{\partial z} + \tilde{U}\frac{\partial \tilde{V}}{\partial x} + \tilde{V}\frac{\partial \tilde{V}}{\partial y} + \tilde{W}\frac{\partial \tilde{V}}{\partial z} =$$

$$-\frac{\partial p}{\partial y} - \frac{\partial \tilde{P}}{\partial y} + \frac{1}{Re}\left(\frac{\partial^2 v}{\partial x^2} + \frac{\partial^2 v}{\partial y^2} + \frac{\partial^2 v}{\partial z^2} + \frac{\partial^2 \tilde{V}}{\partial x^2} + \frac{\partial^2 \tilde{V}}{\partial y^2} + \frac{\partial^2 \tilde{V}}{\partial z^2} \right)$$

$$\frac{\partial w}{\partial t} + \frac{\partial \tilde{W}}{\partial t} + u\frac{\partial w}{\partial x} + u\frac{\partial \tilde{W}}{\partial x} + \tilde{U}\frac{\partial w}{\partial x} + v\frac{\partial w}{\partial y} + v\frac{\partial \tilde{W}}{\partial y} + \tilde{V}\frac{\partial w}{\partial y}$$

$$+ w\frac{\partial w}{\partial z} + w\frac{\partial \tilde{W}}{\partial z} + \tilde{W}\frac{\partial w}{\partial z} + \tilde{U}\frac{\partial \tilde{W}}{\partial x} + \tilde{V}\frac{\partial \tilde{W}}{\partial y} + \tilde{W}\frac{\partial \tilde{W}}{\partial z} =$$

$$-\frac{\partial p}{\partial z} - \frac{\partial \tilde{P}}{\partial z} + \frac{1}{Re}\left(\frac{\partial^2 w}{\partial x^2} + \frac{\partial^2 w}{\partial y^2} + \frac{\partial^2 w}{\partial z^2} + \frac{\partial^2 \tilde{W}}{\partial x^2} + \frac{\partial^2 \tilde{W}}{\partial y^2} + \frac{\partial^2 \tilde{W}}{\partial z^2} \right).$$

In view of (2.20)–(2.24), we get the perturbation equations

$$\frac{\partial u}{\partial x} + \frac{\partial v}{\partial y} + \frac{\partial w}{\partial z} = 0 \tag{2.25}$$

$$\frac{\partial u}{\partial t} + u\frac{\partial u}{\partial x} + u\frac{\partial \tilde{U}}{\partial x} + \tilde{U}\frac{\partial u}{\partial x} + v\frac{\partial u}{\partial y} + v\frac{\partial \tilde{U}}{\partial y} + \tilde{V}\frac{\partial u}{\partial y} + w\frac{\partial u}{\partial z} + w\frac{\partial \tilde{U}}{\partial z} + \tilde{W}\frac{\partial u}{\partial z} =$$

$$-\frac{\partial p}{\partial x} + \frac{1}{Re}\left(\frac{\partial^2 u}{\partial x^2} + \frac{\partial^2 u}{\partial y^2} + \frac{\partial^2 u}{\partial z^2} \right) \tag{2.26}$$

$$\frac{\partial v}{\partial t} + u\frac{\partial v}{\partial x} + \tilde{U}\frac{\partial v}{\partial x} + \frac{\partial \tilde{V}}{\partial x}u + v\frac{\partial v}{\partial y} + v\frac{\partial \tilde{V}}{\partial y} + \tilde{V}\frac{\partial v}{\partial y} + w\frac{\partial v}{\partial z} + w\frac{\partial \tilde{V}}{\partial z} + \tilde{W}\frac{\partial v}{\partial z} =$$

$$-\frac{\partial p}{\partial y} + \frac{1}{Re}\left(\frac{\partial^2 v}{\partial x^2} + \frac{\partial^2 v}{\partial y^2} + \frac{\partial^2 v}{\partial z^2} \right) \tag{2.27}$$

$$\frac{\partial w}{\partial t} + u\frac{\partial w}{\partial x} + u\frac{\partial \tilde{W}}{\partial x} + \tilde{U}\frac{\partial w}{\partial x} + v\frac{\partial w}{\partial y} + v\frac{\partial \tilde{W}}{\partial y} + \tilde{V}\frac{\partial w}{\partial y} + w\frac{\partial w}{\partial z} + w\frac{\partial \tilde{W}}{\partial z} + \tilde{W}\frac{\partial w}{\partial z} =$$
$$-\frac{\partial p}{\partial z} + \frac{1}{Re}\left(\frac{\partial^2 w}{\partial x^2} + \frac{\partial^2 w}{\partial y^2} + \frac{\partial^2 w}{\partial z^2}\right). \quad (2.28)$$

The linearized equations are now obtained by omitting terms that are second order in the perturbations. Thus, we get

$$\frac{\partial u}{\partial x} + \frac{\partial v}{\partial y} + \frac{\partial w}{\partial z} = 0 \quad (2.29)$$

$$\frac{\partial u}{\partial t} + u\frac{\partial \tilde{U}}{\partial x} + \tilde{U}\frac{\partial u}{\partial x} + v\frac{\partial \tilde{U}}{\partial y} + \tilde{V}\frac{\partial u}{\partial y} + w\frac{\partial \tilde{U}}{\partial z} + \tilde{W}\frac{\partial u}{\partial z} =$$
$$-\frac{\partial p}{\partial x} + \frac{1}{Re}\left(\frac{\partial^2 u}{\partial x^2} + \frac{\partial^2 u}{\partial y^2} + \frac{\partial^2 u}{\partial z^2}\right) \quad (2.30)$$

$$\frac{\partial v}{\partial t} + \tilde{U}\frac{\partial v}{\partial x} + \frac{\partial \tilde{V}}{\partial x}u + v\frac{\partial \tilde{V}}{\partial y} + \tilde{V}\frac{\partial v}{\partial y} + w\frac{\partial \tilde{V}}{\partial z} + \tilde{W}\frac{\partial v}{\partial z} =$$
$$-\frac{\partial p}{\partial y} + \frac{1}{Re}\left(\frac{\partial^2 v}{\partial x^2} + \frac{\partial^2 v}{\partial y^2} + \frac{\partial^2 v}{\partial z^2}\right) \quad (2.31)$$

$$\frac{\partial w}{\partial t} + u\frac{\partial \tilde{W}}{\partial x} + \tilde{U}\frac{\partial w}{\partial x} + v\frac{\partial \tilde{W}}{\partial y} + \tilde{V}\frac{\partial w}{\partial y} + w\frac{\partial \tilde{W}}{\partial z} + \tilde{W}\frac{\partial w}{\partial z} =$$
$$-\frac{\partial p}{\partial z} + \frac{1}{Re}\left(\frac{\partial^2 w}{\partial x^2} + \frac{\partial^2 w}{\partial y^2} + \frac{\partial^2 w}{\partial z^2}\right). \quad (2.32)$$

2.5.2   Cylindrical Coordinates

Suppose $(\tilde{V}_r, \tilde{V}_\theta, \tilde{V}_z, \tilde{P})$ is a steady state solution of (2.16) and (2.17)–(2.19). Defining the perturbation

$$v_r \triangleq V_r - \tilde{V}_r$$
$$v_\theta \triangleq V_\theta - \tilde{V}_\theta$$
$$v_z \triangleq V_z - \tilde{V}_z$$
$$p \triangleq P - \tilde{P}$$

we get the perturbation equations

$$\frac{1}{r}\frac{\partial}{\partial r}(rv_r) + \frac{1}{r}\frac{\partial v_\theta}{\partial \theta} + \frac{\partial v_z}{\partial z} = 0 \quad (2.33)$$

$$\frac{\partial v_r}{\partial t} + v_r \frac{\partial v_r}{\partial r} + v_r \frac{\partial \tilde{V}_r}{\partial r} + \tilde{V}_r \frac{\partial v_r}{\partial r} + \frac{v_\theta}{r} \frac{\partial v_r}{\partial \theta} + \frac{v_\theta}{r} \frac{\partial \tilde{V}_r}{\partial \theta} + \frac{\tilde{V}_\theta}{r} \frac{\partial v_r}{\partial \theta}$$

$$- \frac{v_\theta^2}{r} - 2 \frac{v_\theta}{r} \tilde{V}_\theta + v_z \frac{\partial v_r}{\partial z} + v_z \frac{\partial \tilde{V}_r}{\partial z} + \tilde{V}_z \frac{\partial v_r}{\partial z} =$$

$$- \frac{\partial p}{\partial r} + \frac{1}{Re} \left(\frac{\partial}{\partial r} \left(\frac{1}{r} \frac{\partial}{\partial r} (r v_r) \right) + \frac{1}{r^2} \frac{\partial^2 v_r}{\partial \theta^2} - \frac{2}{r^2} \frac{\partial v_\theta}{\partial \theta} + \frac{\partial^2 v_r}{\partial z^2} \right) \quad (2.34)$$

$$\frac{\partial v_\theta}{\partial t} + v_r \frac{\partial v_\theta}{\partial r} + v_r \frac{\partial \tilde{V}_\theta}{\partial r} + \tilde{V}_r \frac{\partial v_\theta}{\partial r} + \frac{v_\theta}{r} \frac{\partial v_\theta}{\partial \theta} + \frac{v_\theta}{r} \frac{\partial \tilde{V}_\theta}{\partial \theta} + \frac{\tilde{V}_\theta}{r} \frac{\partial v_\theta}{\partial \theta}$$

$$+ \frac{v_r v_\theta}{r} + \frac{v_r \tilde{V}_\theta}{r} + \frac{\tilde{V}_r v_\theta}{r} + v_z \frac{\partial v_\theta}{\partial z} + v_z \frac{\partial \tilde{V}_\theta}{\partial z} + \tilde{V}_z \frac{\partial v_\theta}{\partial z} =$$

$$- \frac{1}{r} \frac{\partial p}{\partial \theta} + \frac{1}{Re} \left(\frac{\partial}{\partial r} \left(\frac{1}{r} \frac{\partial}{\partial r} (r v_\theta) \right) + \frac{1}{r^2} \frac{\partial^2 v_\theta}{\partial \theta^2} + \frac{2}{r^2} \frac{\partial v_r}{\partial \theta} + \frac{\partial^2 v_\theta}{\partial z^2} \right) \quad (2.35)$$

$$\frac{\partial v_z}{\partial t} + v_r \frac{\partial v_z}{\partial r} + v_r \frac{\partial \tilde{V}_z}{\partial r} + \tilde{V}_r \frac{\partial v_z}{\partial r} + \frac{v_\theta}{r} \frac{\partial v_z}{\partial \theta} + \frac{v_\theta}{r} \frac{\partial \tilde{V}_z}{\partial \theta} + \frac{\tilde{V}_\theta}{r} \frac{\partial v_z}{\partial \theta}$$

$$+ v_z \frac{\partial v_z}{\partial z} + v_z \frac{\partial \tilde{V}_z}{\partial z} + \tilde{V}_z \frac{\partial v_z}{\partial z} =$$

$$- \frac{\partial p}{\partial z} + \frac{1}{Re} \left(\frac{1}{r} \frac{\partial}{\partial r} \left(r \frac{\partial v_z}{\partial r} \right) + \frac{1}{r^2} \frac{\partial^2 v_z}{\partial \theta^2} + \frac{\partial^2 v_z}{\partial z^2} \right) \quad (2.36)$$

and the linearized equations

$$\frac{1}{r} \frac{\partial}{\partial r} (r v_r) + \frac{1}{r} \frac{\partial v_\theta}{\partial \theta} + \frac{\partial v_z}{\partial z} = 0 \quad (2.37)$$

$$\frac{\partial v_r}{\partial t} + v_r \frac{\partial \tilde{V}_r}{\partial r} + \tilde{V}_r \frac{\partial v_r}{\partial r} + \frac{v_\theta}{r} \frac{\partial \tilde{V}_r}{\partial \theta} + \frac{\tilde{V}_\theta}{r} \frac{\partial v_r}{\partial \theta} - 2 \frac{v_\theta}{r} \tilde{V}_\theta + v_z \frac{\partial \tilde{V}_r}{\partial z} + \tilde{V}_z \frac{\partial v_r}{\partial z} =$$

$$- \frac{\partial p}{\partial r} + \frac{1}{Re} \left(\frac{\partial}{\partial r} \left(\frac{1}{r} \frac{\partial}{\partial r} (r v_r) \right) + \frac{1}{r^2} \frac{\partial^2 v_r}{\partial \theta^2} - \frac{2}{r^2} \frac{\partial v_\theta}{\partial \theta} + \frac{\partial^2 v_r}{\partial z^2} \right) \quad (2.38)$$

$$\frac{\partial v_\theta}{\partial t} + v_r \frac{\partial \tilde{V}_\theta}{\partial r} + \tilde{V}_r \frac{\partial v_\theta}{\partial r} + \frac{v_\theta}{r} \frac{\partial \tilde{V}_\theta}{\partial \theta} + \frac{\tilde{V}_\theta}{r} \frac{\partial v_\theta}{\partial \theta} + \frac{v_r \tilde{V}_\theta}{r} + \frac{\tilde{V}_r v_\theta}{r} + v_z \frac{\partial \tilde{V}_\theta}{\partial z} + \tilde{V}_z \frac{\partial v_\theta}{\partial z} =$$

$$- \frac{1}{r} \frac{\partial p}{\partial \theta} + \frac{1}{Re} \left(\frac{\partial}{\partial r} \left(\frac{1}{r} \frac{\partial}{\partial r} (r v_\theta) \right) + \frac{1}{r^2} \frac{\partial^2 v_\theta}{\partial \theta^2} + \frac{2}{r^2} \frac{\partial v_r}{\partial \theta} + \frac{\partial^2 v_\theta}{\partial z^2} \right) \quad (2.39)$$

$$\frac{\partial v_z}{\partial t} + v_r \frac{\partial \tilde{V}_z}{\partial r} + \tilde{V}_r \frac{\partial v_z}{\partial r} + \frac{v_\theta}{r} \frac{\partial \tilde{V}_z}{\partial \theta} + \frac{\tilde{V}_\theta}{r} \frac{\partial v_z}{\partial \theta} + v_z \frac{\partial \tilde{V}_z}{\partial z} + \tilde{V}_z \frac{\partial v_z}{\partial z} =$$

$$- \frac{\partial p}{\partial z} + \frac{1}{Re} \left(\frac{1}{r} \frac{\partial}{\partial r} \left(r \frac{\partial v_z}{\partial r} \right) + \frac{1}{r^2} \frac{\partial^2 v_z}{\partial \theta^2} + \frac{\partial^2 v_z}{\partial z^2} \right). \quad (2.40)$$

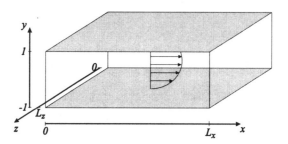

Figure 2.3: Geometry of the 3D channel flow.

2.6   Prototype Flows

So far, the domain on which the Navier-Stokes equation is defined has not entered the picture. In this monograph, the focus is on three prototype flows that have been studied quite extensively in the literature: 1) *channel flow*, which is the flow contained between two parallel plates; 2) *pipe flow*, which is the flow contained in the interior of a cylinder with circular cross section, and; 3) *cylinder flow*, which is the flow past a cylinder with circular cross section. The particularly simple geometries of these flows make the problems mathematically tractable, as well as numerically feasible. In addition, the pipe is encountered frequently in practice, so understanding how to control this flow is of great importance in engineering applications.

2.6.1   3D Channel Flow

The domain in this case is the box $\Omega = \{(x, y, z) \in [0, L_x) \times (-1, 1) \times [0, L_z)\}$, which is shown in Figure 2.3. The flow problem is completed by specifying boundary conditions on the boundary of the box. In the streamwise (x) and spanwise (z) directions, we use periodic boundary conditions. That is, we equate the flow quantities at $x = 0$ and $x = L_x$, and at $z = 0$ and $z = L_z$. At the walls, the boundary conditions will eventually be specified by the result of some boundary control design, but for the time being, we select no-slip[1] boundary conditions, i.e. $U = V = W = 0$ at the walls ($y = \pm 1$). Given the boundary conditions, we are now in the position to solve (2.12)–(2.15) for a steady-state solution. In solving the Navier-Stokes equations for a steady-state solution in channel flow with no-slip boundary conditions at the walls, we assume that the velocity field is independent of the streamwise and spanwise directions. Thus, setting the time derivatives and the spatial derivatives with respect to x and z

---

[1]Throughout this text, *no-slip* boundary condition implies no-slip/non-penetration. That is, all components of the velocity are zero.

in (2.12)–(2.15) to zero, we get

$$\frac{\partial V}{\partial y} = 0 \qquad (2.41)$$

$$V\frac{\partial U}{\partial y} = -\frac{\partial P}{\partial x} + \frac{1}{Re}\frac{\partial^2 U}{\partial y^2} \qquad (2.42)$$

$$V\frac{\partial V}{\partial y} = -\frac{\partial P}{\partial y} + \frac{1}{Re}\frac{\partial^2 V}{\partial y^2} \qquad (2.43)$$

$$V\frac{\partial W}{\partial y} = -\frac{\partial P}{\partial z} + \frac{1}{Re}\frac{\partial^2 W}{\partial y^2}. \qquad (2.44)$$

From (2.41) we have that $V = constant$, and since we are employing no-slip boundary conditions, we obtain $V = 0$. So, from (2.42)–(2.44) we have that

$$\frac{\partial P}{\partial x} = \frac{1}{Re}\frac{\partial^2 U}{\partial y^2} \qquad (2.45)$$

$$\frac{\partial P}{\partial y} = 0 \qquad (2.46)$$

$$\frac{\partial P}{\partial z} = \frac{1}{Re}\frac{\partial^2 W}{\partial y^2}. \qquad (2.47)$$

From (2.46), we have that $P = P(x, z)$, and since $U = U(y)$ by assumption, we get

$$\frac{\partial P}{\partial x} = c_1 = \frac{1}{Re}\frac{\partial^2 U}{\partial y^2} \qquad (2.48)$$

$$\frac{\partial P}{\partial z} = c_2 = \frac{1}{Re}\frac{\partial^2 W}{\partial y^2} \qquad (2.49)$$

where c_1 and c_2 are constants. Thus, we can solve for each side of (2.48) and (2.49) separately, to get

$$P = c_1 x + c_2 z + c_3$$

where c_3 is a constant, and

$$U(y) = \frac{c_1 Re}{2}y^2 + c_4 y + c_5$$

$$W(y) = \frac{c_2 Re}{2}y^2 + c_6 y + c_7$$

where c_4, c_5, c_6 and c_7 are constants. Since only the gradient of P enters the Navier-Stokes equations, c_3 may be arbitrarily chosen, so we set $c_3 = 0$. By employing no-slip boundary conditions we obtain the set of equations

$$\frac{c_1 Re}{2} - c_4 + c_5 = 0$$

$$\frac{c_1 Re}{2} + c_4 + c_5 = 0$$

$$\frac{c_2 Re}{2} - c_6 + c_7 = 0$$

$$\frac{c_2 Re}{2} + c_6 + c_7 = 0$$

so that

$$c_4 = c_6 = 0$$

$$c_5 = -\frac{c_1 Re}{2}$$

$$c_7 = -\frac{c_2 Re}{2}.$$

Without loss of generality we select the direction of flow along the positive x-axis, and the center velocity, $U(0) = 1$. We get

$$c_1 = -\frac{2}{Re}, \quad c_2 = 0, \ c_5 = 1, \text{ and } c_7 = 0.$$

So we have the steady-state solution

$$\left(\tilde{U}, \tilde{V}, \tilde{W}, \tilde{P}\right) = \left(1 - y^2, 0, 0, -\frac{2}{Re}x\right). \tag{2.50}$$

Inserting (2.50) into the perturbation equations (2.25)–(2.28) yields

$$\frac{\partial u}{\partial x} + \frac{\partial v}{\partial y} + \frac{\partial w}{\partial z} = 0 \tag{2.51}$$

$$\frac{\partial u}{\partial t} + u\frac{\partial u}{\partial x} + \tilde{U}\frac{\partial u}{\partial x} + v\frac{\partial u}{\partial y} + v\frac{\partial \tilde{U}}{\partial y} + w\frac{\partial u}{\partial z} =$$
$$-\frac{\partial p}{\partial x} + \frac{1}{Re}\left(\frac{\partial^2 u}{\partial x^2} + \frac{\partial^2 u}{\partial y^2} + \frac{\partial^2 u}{\partial z^2}\right) \tag{2.52}$$

$$\frac{\partial v}{\partial t} + u\frac{\partial v}{\partial x} + \tilde{U}\frac{\partial v}{\partial x} + v\frac{\partial v}{\partial y} + w\frac{\partial v}{\partial z} =$$
$$-\frac{\partial p}{\partial y} + \frac{1}{Re}\left(\frac{\partial^2 v}{\partial x^2} + \frac{\partial^2 v}{\partial y^2} + \frac{\partial^2 v}{\partial z^2}\right) \tag{2.53}$$

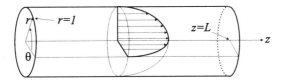

Figure 2.4: Geometry of the pipe flow.

$$\frac{\partial w}{\partial t} + u\frac{\partial w}{\partial x} + \tilde{U}\frac{\partial w}{\partial x} + v\frac{\partial w}{\partial y} + w\frac{\partial w}{\partial z} =$$
$$- \frac{\partial p}{\partial z} + \frac{1}{Re}\left(\frac{\partial^2 w}{\partial x^2} + \frac{\partial^2 w}{\partial y^2} + \frac{\partial^2 w}{\partial z^2}\right). \quad (2.54)$$

The linearized equations are

$$\frac{\partial u}{\partial x} + \frac{\partial v}{\partial y} + \frac{\partial w}{\partial z} = 0 \quad (2.55)$$

$$\frac{\partial u}{\partial t} + \tilde{U}\frac{\partial u}{\partial x} + v\frac{\partial \tilde{U}}{\partial y} = -\frac{\partial p}{\partial x} + \frac{1}{Re}\left(\frac{\partial^2 u}{\partial x^2} + \frac{\partial^2 u}{\partial y^2} + \frac{\partial^2 u}{\partial z^2}\right) \quad (2.56)$$

$$\frac{\partial v}{\partial t} + \tilde{U}\frac{\partial v}{\partial x} = -\frac{\partial p}{\partial y} + \frac{1}{Re}\left(\frac{\partial^2 v}{\partial x^2} + \frac{\partial^2 v}{\partial y^2} + \frac{\partial^2 v}{\partial z^2}\right) \quad (2.57)$$

$$\frac{\partial w}{\partial t} + \tilde{U}\frac{\partial w}{\partial x} = -\frac{\partial p}{\partial z} + \frac{1}{Re}\left(\frac{\partial^2 w}{\partial x^2} + \frac{\partial^2 w}{\partial y^2} + \frac{\partial^2 w}{\partial z^2}\right). \quad (2.58)$$

Choosing the channel half width (which is 1) as the characteristic length, and the center velocity (which is 1) as the characteristic velocity, the Reynolds number is simply $Re = \rho/\mu$.

2.6.2 3D Pipe Flow

The domain in this case is the cylinder $\Omega = \{(r, \theta, z) \in [0, 1) \times [0, 2\pi) \times [0, L)\}$, which is shown in Figure 2.4. In the angular (θ) direction the boundary conditions are clearly periodic. In the streamwise (z) direction, we also use periodic boundary conditions. That is, we equate the flow quantities at $\theta = 0$ and $\theta = 2\pi$, and at $z = 0$ and $z = L$. In the radial direction (r) we impose the boundary conditions that the velocity be finite at $r = 0$, and no-slip at the wall $(r = 1)$. We are now in the position to solve (2.16)–(2.19) for a steady-state solution, which we assume is independent of the angular (θ) and streamwise (z) directions with $V_r = V_\theta = 0$. Thus, we get

$$0 = \frac{\partial P}{\partial r} \tag{2.59}$$

$$0 = \frac{\partial P}{\partial \theta} \tag{2.60}$$

$$\frac{\partial P}{\partial z} = \frac{1}{Re} \frac{1}{r} \frac{\partial}{\partial r} \left(r \frac{\partial V_z}{\partial r} \right). \tag{2.61}$$

From (2.59) and (2.60) we have that $P = P(z)$, and since $V_z = V_z(r)$ by assumption, we get

$$\frac{\partial P}{\partial z} = -c = \frac{1}{Re} \frac{1}{r} \frac{\partial}{\partial r} \left(r \frac{\partial V_z}{\partial r} \right) \tag{2.62}$$

where c is a constant. The left hand side of (2.62) can be integrated to obtain

$$P = -cz + c_1$$

and the right hand side of (2.62) is separable, so we can solve it to obtain

$$V_z(r) = -\frac{cRe}{4} r^2 + c_2 \ln r + c_3.$$

Imposing the boundary conditions in the radial direction, yields

$$c_2 = 0 \text{ and } c_3 = \frac{cRe}{4}$$

and without loss of generality we select the center velocity, $V_z(0) = 1$, so that $c = \frac{4}{Re}$. Thus, we have the steady-state solution

$$\left(\tilde{V}_r, \tilde{V}_\theta, \tilde{V}_z, \tilde{P} \right) = \left(0, 0, 1 - r^2, -\frac{4}{Re} z \right). \tag{2.63}$$

The perturbation equations are

$$\frac{1}{r} \frac{\partial}{\partial r} (r v_r) + \frac{1}{r} \frac{\partial v_\theta}{\partial \theta} + \frac{\partial v_z}{\partial z} = 0 \tag{2.64}$$

$$\frac{\partial v_r}{\partial t} + v_r \frac{\partial v_r}{\partial r} + \frac{v_\theta}{r} \frac{\partial v_r}{\partial \theta} - \frac{v_\theta^2}{r} + v_z \frac{\partial v_r}{\partial z} + \tilde{V}_z \frac{\partial v_r}{\partial z} =$$
$$- \frac{\partial p}{\partial r} + \frac{1}{Re} \left(\frac{\partial}{\partial r} \left(\frac{1}{r} \frac{\partial}{\partial r} (r v_r) \right) + \frac{1}{r^2} \frac{\partial^2 v_r}{\partial \theta^2} - \frac{2}{r^2} \frac{\partial v_\theta}{\partial \theta} + \frac{\partial^2 v_r}{\partial z^2} \right) \tag{2.65}$$

$$\frac{\partial v_\theta}{\partial t} + v_r \frac{\partial v_\theta}{\partial r} + \frac{v_\theta}{r} \frac{\partial v_\theta}{\partial \theta} + \frac{v_r v_\theta}{r} + v_z \frac{\partial v_\theta}{\partial z} + \tilde{V}_z \frac{\partial v_\theta}{\partial z} =$$
$$- \frac{1}{r} \frac{\partial p}{\partial \theta} + \frac{1}{Re} \left(\frac{\partial}{\partial r} \left(\frac{1}{r} \frac{\partial}{\partial r} (r v_\theta) \right) + \frac{1}{r^2} \frac{\partial^2 v_\theta}{\partial \theta^2} + \frac{2}{r^2} \frac{\partial v_r}{\partial \theta} + \frac{\partial^2 v_\theta}{\partial z^2} \right) \tag{2.66}$$

$$\frac{\partial v_z}{\partial t} + v_r \frac{\partial v_z}{\partial r} + v_r \frac{\partial \tilde{V}_z}{\partial r} + \frac{v_\theta}{r} \frac{\partial v_z}{\partial \theta} + v_z \frac{\partial v_z}{\partial z} + \tilde{V}_z \frac{\partial v_z}{\partial z} =$$
$$-\frac{\partial p}{\partial z} + \frac{1}{Re} \left(\frac{1}{r} \frac{\partial}{\partial r} \left(r \frac{\partial v_z}{\partial r} \right) + \frac{1}{r^2} \frac{\partial^2 v_z}{\partial \theta^2} + \frac{\partial^2 v_z}{\partial z^2} \right) \quad (2.67)$$

and the linearized equations are

$$\frac{1}{r} \frac{\partial}{\partial r} (r v_r) + \frac{1}{r} \frac{\partial v_\theta}{\partial \theta} + \frac{\partial v_z}{\partial z} = 0 \quad (2.68)$$

$$\frac{\partial v_r}{\partial t} + \tilde{V}_z \frac{\partial v_r}{\partial z} =$$
$$-\frac{\partial p}{\partial r} + \frac{1}{Re} \left(\frac{\partial}{\partial r} \left(\frac{1}{r} \frac{\partial}{\partial r} (r v_r) \right) + \frac{1}{r^2} \frac{\partial^2 v_r}{\partial \theta^2} - \frac{2}{r^2} \frac{\partial v_\theta}{\partial \theta} + \frac{\partial^2 v_r}{\partial z^2} \right) \quad (2.69)$$

$$\frac{\partial v_\theta}{\partial t} + \tilde{V}_z \frac{\partial v_\theta}{\partial z} =$$
$$-\frac{1}{r} \frac{\partial p}{\partial \theta} + \frac{1}{Re} \left(\frac{\partial}{\partial r} \left(\frac{1}{r} \frac{\partial}{\partial r} (r v_\theta) \right) + \frac{1}{r^2} \frac{\partial^2 v_\theta}{\partial \theta^2} + \frac{2}{r^2} \frac{\partial v_r}{\partial \theta} + \frac{\partial^2 v_\theta}{\partial z^2} \right) \quad (2.70)$$

$$\frac{\partial v_z}{\partial t} + v_r \frac{\partial \tilde{V}_z}{\partial r} + \tilde{V}_z \frac{\partial v_z}{\partial z} =$$
$$-\frac{\partial p}{\partial z} + \frac{1}{Re} \left(\frac{1}{r} \frac{\partial}{\partial r} \left(r \frac{\partial v_z}{\partial r} \right) + \frac{1}{r^2} \frac{\partial^2 v_z}{\partial \theta^2} + \frac{\partial^2 v_z}{\partial z^2} \right). \quad (2.71)$$

Choosing the pipe radius (which is 1) as the characteristic length, and the center velocity (which is 1) as the characteristic velocity, the Reynolds number is simply $Re = \rho/\mu$.

2.6.3 2D Channel/Pipe Flow

In 2D, channel and pipe flows have the same domain, which is shown in Figure 2.5. The domain is in this case $\Omega = \{(x, y) \in [0, L] \times (-1, 1)\}$, and periodic boundary conditions are employed in the streamwise (x) direction. With no-slip boundary conditions at the wall, we have the steady-state solution

$$\left(\tilde{U}, \tilde{V}, \tilde{P} \right) = \left(1 - y^2, 0, -\frac{2}{Re} x \right) \quad (2.72)$$

and the perturbation equations become

$$\frac{\partial u}{\partial x} + \frac{\partial v}{\partial y} = 0 \quad (2.73)$$

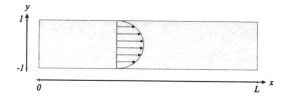

Figure 2.5: Geometry of the 2D channel flow.

$$\frac{\partial u}{\partial t} + u\frac{\partial u}{\partial x} + \tilde{U}\frac{\partial u}{\partial x} + v\frac{\partial u}{\partial y} + v\frac{\partial \tilde{U}}{\partial y} = -\frac{\partial p}{\partial x} + \frac{1}{Re}\left(\frac{\partial^2 u}{\partial x^2} + \frac{\partial^2 u}{\partial y^2}\right) \quad (2.74)$$

$$\frac{\partial v}{\partial t} + u\frac{\partial v}{\partial x} + \tilde{U}\frac{\partial v}{\partial x} + v\frac{\partial v}{\partial y} = -\frac{\partial p}{\partial y} + \frac{1}{Re}\left(\frac{\partial^2 v}{\partial x^2} + \frac{\partial^2 v}{\partial y^2}\right). \quad (2.75)$$

The linearized equations are

$$\frac{\partial u}{\partial x} + \frac{\partial v}{\partial y} = 0 \quad (2.76)$$

$$\frac{\partial u}{\partial t} + \tilde{U}\frac{\partial u}{\partial x} + v\frac{\partial \tilde{U}}{\partial y} = -\frac{\partial p}{\partial x} + \frac{1}{Re}\left(\frac{\partial^2 u}{\partial x^2} + \frac{\partial^2 u}{\partial y^2}\right) \quad (2.77)$$

$$\frac{\partial v}{\partial t} + \tilde{U}\frac{\partial v}{\partial x} = -\frac{\partial p}{\partial y} + \frac{1}{Re}\left(\frac{\partial^2 v}{\partial x^2} + \frac{\partial^2 v}{\partial y^2}\right). \quad (2.78)$$

Choosing the channel half width (which is 1) as the characteristic length, and the center velocity (which is 1) as the characteristic velocity, the Reynolds number is simply $Re = \rho/\mu$.

2.6.4   2D Cylinder Flow

For the 2D cylinder flow, the domain is the entire plane, except for the unit disc (see Figure 2.6). That is, $\Omega = \{(r,\theta) \in (1,\infty) \times [0,2\pi)\}$. In the angular (θ) direction the boundary conditions are periodic, and in the radial (r) direction we impose no-slip at the wall $(r = 1)$, and uniform flow with unit amplitude at $r = \infty$, that is, $V_r(\infty,\theta) = cos(\theta)$ and $V_\theta(\infty,\theta) = -sin(\theta)$. Choosing the disc diameter (which is 2) as the characteristic length, and the velocity at infinity (which is 1) as the characteristic velocity, the Reynolds number is simply $Re = 2\rho/\mu$. In great contrast to the previous cases, we are unable to solve for a steady-state solution of the cylinder flow. Consequently, in what follows, this case is treated numerically, only. However, a simplified model of vortex shedding, the Ginzburg-Landau model, is presented and analyzed with respect to stability in Section 4.4.

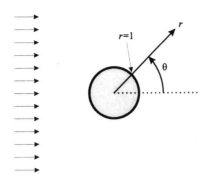

Figure 2.6: Geometry of the 2D cylinder flow.

2.7 Spatial Discretization

For simulation purposes, as well as some approaches to control design, one needs to discretize the equations of motion both spatially and temporally. A general treatment of discretization methods for the Navier-Stokes equation is given in [49]. In this section, the treatment of this subject is restricted to the methods used in the design of controllers in the next chapter. These are so-called spectral methods, which lend themselves particularly well suited for the prototype flows that were described in the previous section. In [35], an in-depth treatment of spectral methods applied to partial differential equations, including the two methods outlined below, is given.

2.7.1 Spectral Methods

Consider the evolution equation

$$\frac{\partial u}{\partial t} = A(u) \tag{2.79}$$

with the initial condition

$$u(x, 0) = u_0(x),$$

where $u \in \mathcal{X}$ is the solution sought, and A is an operator containing the spatial derivatives of u. The equation is defined on the spatial domain Ω, which we will assume is one-dimensional for simplicity, that is, $x \in \Omega \subset \mathbb{R}$. $\mathcal{X}$ is a Hilbert space with scalar product

$$(u, v)_{\mathcal{X}} = \int_{\Omega} u\bar{v} d\Omega$$

and norm

$$\|u\|_{\mathcal{X}} = (u, u)^{\frac{1}{2}}.$$

The bar ($\bar{v}$) denotes complex conjugation. In order to complete the problem, appropriate boundary conditions must be supplied. Now, consider a series expansion for the solution u, that is

$$u^*(x, t) = \sum_{k=0}^{\infty} a_k(t)\phi_k(x) \tag{2.80}$$

where the ϕ_k are called the *trial functions*, and the a_k are called the *expansion coefficients*. The ϕ_k are assumed to constitute a complete basis for $\mathcal{X}$, so that the series (2.80) converges to u in $\mathcal{X}$. The approximation of (2.80), of order N, is defined as the truncated series

$$u^N(x, t) = \sum_{k \in I} a_k(t)\phi_k(x),$$

where I is a finite set of indices. Thus, $u^N \in \mathcal{X}^N = span\{\phi_k \,|\, k \in I\} \subset \mathcal{X}$. Since u^N is an approximate solution of (2.79), the residual

$$R\left(u^N\right) = \frac{\partial u^N}{\partial t} - A\left(u^N\right)$$

will in general not vanish everywhere, but its projection onto the span of a set of *test functions*, $\mathcal{Y}^N = span\{\psi_j \,|\, j \in I\}$, is required to be zero. That is,

$$\left(\frac{\partial u^N}{\partial t} - A(u^N), \psi_j\right)_{\mathcal{X}} = \int_{\Omega} \left(\frac{\partial u^N}{\partial t} - A(u^N)\right) \bar{\psi}_j d\Omega = 0, \ j \in I. \tag{2.81}$$

Equation (2.81) results in N ordinary differential equation for the determination of the expansion coefficients a_k. The finite set of trial functions define the space in which to search for a solution to the reduced problem, whereas the test functions define a space in which the projection of the residual must vanish. The two spaces may or may not be the same, as the next two sections demonstrate.

2.7.2 The Fourier-Galerkin Method

The Galerkin method is characterized by the fact that the trial functions and the test functions are the same, and that each test function satisfies the boundary conditions. When the boundary conditions are periodic with period L, that is $\Omega = [0, L)$, a natural choice for the trial functions is

$$\phi_k = \psi_k = e^{ik\frac{2\pi}{L}x}, \ k = -N/2, ..., N/2.$$

With this choice of test and trial functions, equation (2.81) is called the Fourier-Galerkin method. The ϕ_k constitute an orthogonal basis for $\mathcal{X}^N$, since

$$(\phi_j, \phi_k)_{\mathcal{X}} = \int_0^L \phi_j \bar{\phi}_k dx = \int_0^L e^{i(j-k)\frac{2\pi}{L}x} dx = \left\{ \begin{array}{l} L, \; j = k \\ 0, \; j \neq k \end{array} \right. .$$

The main advantage of using trigonometric polynomials is the simplicity and accuracy in calculating the derivatives

$$\frac{d^m}{dx^m}\phi_k = \left(\frac{2\pi i k}{L} \right)^m \phi_k. \tag{2.82}$$

As an example, assume that the operator A is linear, and contains spatial derivatives up to order M. In view of (2.82), the operator A acting on ϕ_k can be represented as the finite sum

$$A(\phi_k) = \sum_{m=0}^{M} c_{mk} \phi_k$$

where the c_{mk} are complex constants. From (2.81), keeping in mind that A is linear and that the scalar product is bilinear, we get

$$\left(\frac{\partial}{\partial t} \left(\sum_{k=-N/2}^{N/2} a_k(t)\phi_k \right) - A \left(\sum_{k=-N/2}^{N/2} a_k(t)\phi_k \right), \phi_j \right)_{\mathcal{X}} =$$

$$\sum_{k=-N/2}^{N/2} \left(\frac{\partial a_k(t)}{\partial t} - a_k(t) \sum_{m=0}^{M} c_{mk} \right) (\phi_k, \phi_j)_{\mathcal{X}} = L \left(\frac{\partial a_j(t)}{\partial t} - a_j(t) \sum_{m=0}^{M} c_{mj} \right).$$

Thus, the finite set of ordinary differential equations

$$\frac{da_k(t)}{dt} - a_k(t) \sum_{m=0}^{M} c_{mk} = 0, \text{ for } k = -N/2, ..., N/2$$

determines the expansion coefficients a_k. The initial condition is obtained by the usual relation (Fourier transform)

$$a_k(0) = \frac{1}{L} \int_0^L u_0(x) e^{-ik\frac{2\pi}{L}x} dx, \; k = -N/2, ..., N/2.$$

Due to the periodic trial functions, the Fourier-Galerkin method is a natural choice of discretization method for the streamwise and spanwise directions of channel flows.

2.7.3   The Chebyshev Collocation Method

The collocation method is characterized by the fact that the test functions are shifted *Dirac* functions, $\psi_j(x) = \delta(x - x_j)$, which are defined by

$$\int_\Omega \delta\left(x - x_j\right) f dx = f(x_j).$$

Thus, (2.81) reduces to

$$\left(\frac{\partial u^N}{\partial t} - A(u^N), \psi_j\right)_\chi = \int_\Omega \left(\frac{\partial u^N}{\partial t} - A(u^N)\right) \bar{\psi}_j d\Omega =$$

$$\left(\frac{\partial u^N\left(x_j\right)}{\partial t} - A(u^N\left(x_j\right))\right) = 0, \ j = 1, 2, 3, ..., N-1. \quad (2.83)$$

The boundary conditions are taken to be $u^N(-1,t) = u^N(x_0,t)$, and $u^N(1,t) = u^N(x_N,t)$, and the initial condition is $u^N(x_j,0) = u(x_j,0)$, for $j = 0, 1, 2, ..., N$. Unlike the Galerkin method, which is implemented in terms of the expansion coefficients, a_k, the collocation method is implemented in terms of u at the collocation points, x_j. The expansion coefficients are used in the differentiation of u^N. In the Chebyshev collocation method, the trial functions are Chebyshev polynomials (or a linear combination of Chebyshev polynomials), which are defined on $\Omega = [-1, 1]$ as

$$\phi_k(x) = \cos\left(k\theta\right), \ \theta = \cos^{-1}\left(x\right) \text{ for } k = 0, 1, 2, 3, ...$$

The Chebyshev expansion of u is

$$u^*(x) = \sum_{k=0}^\infty a_k \phi_k(x), \ a_k = \frac{2}{\pi c_k} \int_{-1}^1 u(x)\phi_k(x)w(x)dx \quad (2.84)$$

where

$$c_k = \begin{cases} 2, k = 0 \\ 1, k > 0 \end{cases}$$

$$w(x) = \frac{1}{\sqrt{1 - x^2}}.$$

The derivative of (2.84) is

$$\frac{du^*(x)}{dx} = \sum_{k=0}^\infty a_k \frac{d\phi_k(x)}{dx}$$

which is computed most efficiently by deriving a recursive formula. For $k \geq 1$, we have that

$$\frac{d\phi_{k-1}(x)}{dx} = -(k-1)\sin\left((k-1)\theta\right)\frac{d\theta}{dx}$$

$$\frac{d\phi_{k+1}(x)}{dx} = -(k+1)\sin\left((k+1)\theta\right)\frac{d\theta}{dx}$$

which, by standard trigonometric relationships, gives

$$\frac{1}{k+1}\frac{d\phi_{k+1}(x)}{dx} - \frac{1}{k-1}\frac{d\phi_{k-1}(x)}{dx} = -2\cos{(k\theta)}\sin{(\theta)}\frac{d\theta}{dx}.$$

Since

$$\frac{d\theta}{dx} = -\frac{1}{\sqrt{1-x^2}} = -\frac{1}{\sin{(\theta)}}$$

we obtain

$$\frac{1}{k+1}\frac{d\phi_{k+1}(x)}{dx} - \frac{1}{k-1}\frac{d\phi_{k-1}(x)}{dx} = 2\phi_k\left(x\right).$$

Thus, the derivative of the trial functions obeys the recursive formula

$$\frac{d\phi_k(x)}{dx} = \frac{k}{k-2}\frac{d\phi_{k-2}(x)}{dx} + 2k\phi_{k-1}\left(x\right), \ k \geq 2. \qquad (2.85)$$

Repeated use of (2.85) yields

$$\frac{d\phi_k(x)}{dx} = 2k\left(\phi_{k-1}\left(x\right) + \phi_{k-3}\left(x\right) + \phi_{k-5}\left(x\right) + \cdots + \frac{1}{2-l}\phi_l\left(x\right)\right)$$

$$k \geq 2, \ l = \begin{cases} 0 \text{ for } k \text{ odd} \\ 1 \text{ for } k \text{ even} \end{cases}. \qquad (2.86)$$

From $d\phi_0\left(x\right)/dx = 0$, $d\phi_1\left(x\right)/dx = \phi_0(x)$, we see that (2.86) is valid for all $k \geq 0$. It follows from (2.86) that the expansion coefficients for du/dx, denoted a_k^d, are given as

$$a_0^d = \sum_{\substack{m=1 \\ m \text{ odd}}}^{\infty} m a_m \qquad (2.87)$$

$$a_k^d = 2 \sum_{\substack{m=k+1 \\ m+k \text{ odd}}}^{\infty} m a_m, \ k \geq 1. \qquad (2.88)$$

The accuracy of the approximation is highly dependent on the choice of collocation points, and a common choice is (Chebyshev-Gauss-Lobatto points)

$$x_j = -\cos{\frac{\pi j}{N}}, \ j = 0, 1, 2, ..., N. \qquad (2.89)$$

In the discrete case, the expansion of the approximation is the truncated series

$$u^N(x) = \sum_{k=0}^{N} a_k \phi_k(x)$$

with the discrete expansion coefficients

$$a_k = \frac{1}{\gamma_k} \sum_{j=0}^{N} u(x_j)\phi_k(x_j)w_j$$

where

$$\gamma_k = \left\{ \begin{array}{l} \pi, \text{ for } k = 0, N \\ \frac{\pi}{2}, \text{ for } 0 < k < N \end{array} \right. , \text{ and } w_j = \left\{ \begin{array}{l} \frac{\pi}{2N}, \text{ for } j = 0, N \\ \frac{\pi}{N}, \text{ for } 0 < j < N \end{array} \right. .$$

Equations (2.87)–(2.88) provide a recursive formula for the calculation of the derivatives in Chebyshev space

$$a_{k \geq N}^d = 0$$

$$a_k^d = 2 \sum_{\substack{m=k+1 \\ m+k \text{ odd}}}^{N} m a_m = 2(k+1)a_{k+1} + a_{k+2}^d, \ k \geq 1$$

$$a_0^d = a_1 + \sum_{\substack{m=3 \\ m \text{ odd}}}^{\infty} m a_m = a_1 + \frac{1}{2}a_2^d.$$

As for the Fourier-Galerkin method, the result is a finite set of ordinary differential equations, which are difficult to write in this case. The discrete derivative may also be expressed as a matrix multiplication, that is

$$\frac{du^N(x_k)}{dx} = \sum_{j=0}^{N} (\mathcal{D}_N)_{kj} u^N(x_j)$$

where $\mathcal{D}_N$ is the $N+1 \times N+1$ matrix

$$(\mathcal{D}_N)_{kj} \triangleq d_{kj} = \left\{ \begin{array}{ll} \frac{\bar{c}_k}{\bar{c}_j} \frac{(-1)^{k+j}}{x_k - x_j} & k \neq j \\ -\frac{x_j}{2(1-x_j^2)} & 1 \leq k = j \leq N-1 \\ \frac{2N^2+1}{6} & k = j = 1 \\ -\frac{2N^2+1}{6} & k = j = N \end{array} \right. . \tag{2.90}$$

Computationally, the Chebyshev transform is superior to matrix multiplication for computing the derivative when $N >\sim 20$. The Chebyshev collocation method is well suited for discretizing the wall-normal direction in channel flows, since it allows nonuniform distribution of the collocation points. Thus, the boundary layer can be adequately resolved by placing collocation points more densely near the walls.

Chapter 3

Control Theoretic Preliminaries

The following control theoretic preliminaries are based on the books [10, 52, 56, 136, 121].

3.1 Linear Time-Invariant Systems

Consider the following linear time-invariant system

$$\dot{x} = Ax + Bu \tag{3.1}$$
$$y = Cx + Du \tag{3.2}$$

with initial condition $x(t_0) = x_0$. The system transfer function for this system is

$$G(s) = C(sI - A)^{-1}B + D.$$

The fact that $G(s)$ has the state-space realization (3.1)–(3.2) is denoted

$$G(s) = \left[\begin{array}{c|c} A & B \\ \hline C & D \end{array} \right].$$

3.1.1 Classical Control

Here we will review one result from classical control theory, namely the *Nyquist stability criterion*, which is based on the argument principle of complex analysis. In short, the Nyquist stability criterion relates the open-loop frequency response

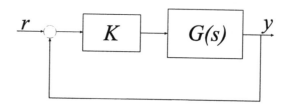

Figure 3.1: Feedback configuration for application of the Nyquist stability criterion.

to the number of poles of the system in the right half of the complex plane. We study the feedback configuration shown schematically in Figure 3.1, and for simplicity, we consider a single-input-single-output (SISO) system and assume that the controller is a simple proportional controller with constant feedback gain. To determine stability of the closed loop system

$$G_c(s) = \frac{KG(s)}{1 + KG(s)}$$

where $G(s)$ is the plant transfer function, and K is the feedback gain, the Nyquist stability criterion can be used in the following three-step procedure

1. Determine the number of unstable poles of $G(s)$ and call that number P.

2. Plot the Nyquist plot of $G(s)$ (which is simply the curve $(real(G(j\omega)), imag(G(j\omega))$ for $\omega \in \mathbb{R}$), and evaluate the number of times the curve encircles $-1/K$ in the clockwise direction. Call that number N.

3. The number of unstable closed-loop roots is $Z = N + P$.

3.1.2 LQG Control

In LQG control (linear-quadratic-gaussian), the system (3.1)–(3.2) is assumed to be subjected to disturbances entering additively in both equations. Thus, we have

$$\begin{aligned}
\dot{x} &= Ax + Bu + w_d & (3.3) \\
y &= Cx + Du + w_n & (3.4)
\end{aligned}$$

where w_d is a disturbance and w_n is measurement noise. w_d and w_n are assumed to be uncorrelated Gaussian stochastic processes with zero means and covariances

$$\begin{aligned}
E\left\{ w_d(t)\, w_d^T(\tau) \right\} &= W\delta(t - \tau), \text{ and} \\
E\left\{ w_n(t)\, w_n^T(\tau) \right\} &= V\delta(t - \tau),
\end{aligned}$$

respectively, where W and V are constant matrices. E denotes the expectation operator, and δ denotes the delta function. The LQG control problem is to find the controller K that minimizes the cost functional

$$J = E \left\{ \lim_{T \to \infty} \frac{1}{T} \int_{t_0}^{T} \left(x^T Q x + u^T R u \right) dt \right\}$$

where Q and R are constant weighting matrices satisfying $Q = Q^T \geq 0$ and $R = R^T > 0$. This problem is solved by a three step procedure: 1) find the optimal state feedback when the noise is ignored; 2) find the optimal state estimator; and 3) replace the state in 1) with its estimate. The fact that the state feedback law and the estimator can be designed independently is referred to as the *separation theorem* (or the *certainty equivalence principle*). The optimal state feedback is the feedback u that minimizes the cost functional

$$J_s = \lim_{T \to \infty} \int_{t_0}^{T} \left(x^T Q x + u^T R u \right) dt.$$

If (A, B) is stabilizable, the problem has a unique solution given by

$$u(t) = -F x(t),$$

where $F = R^{-1} B^T X$ and $X = X^T \geq 0$ uniquely solves the algebraic Riccati equation

$$A^T X + X A - X B R^{-1} B^T X + Q = 0.$$

The optimal state estimator, or the *Kalman filter*, having the structure

$$\dot{\hat{x}} = A \hat{x} + B u + L \left(y - C \hat{x} \right) \tag{3.5}$$

is the one that minimizes the covariance of the estimation error

$$E \left\{ (x - \hat{x})^T (x - \hat{x}) \right\}.$$

If (A, C) is detectable, the problem has a unique solution given by

$$L = Y C^T V^{-1}$$

where $Y = Y^T \geq 0$ uniquely solves the algebraic Riccati equation

$$Y A^T + A Y - Y C^T V^{-1} C Y + W = 0.$$

In summary, the controller K is given as

$$K(s) = \left[\begin{array}{c|c} A - BF - LC & L \\ \hline -F & 0 \end{array} \right]. \tag{3.6}$$

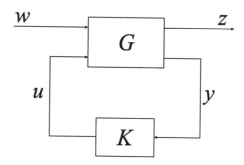

Figure 3.2: Standard control system configuration.

3.1.3 $\mathcal{H}_2$ Control

The standard $\mathcal{H}_2$ control problem considers the system shown in Figure 3.2, which can be described by

$$
\begin{bmatrix} z \\ y \end{bmatrix} = G(s) \begin{bmatrix} w \\ u \end{bmatrix} = \begin{bmatrix} G_{11}(s) & G_{12}(s) \\ G_{21}(s) & G_{22}(s) \end{bmatrix} \begin{bmatrix} w \\ u \end{bmatrix} \tag{3.7}
$$
$$
u = K(s)y
$$

with the state-space realization

$$
G(s) = \left[\begin{array}{c|c} A & \begin{bmatrix} B_1 & B_2 \end{bmatrix} \\ \hline \begin{bmatrix} C_1 \\ C_2 \end{bmatrix} & \begin{bmatrix} 0 & D_{12} \\ D_{21} & 0 \end{bmatrix} \end{array} \right], \tag{3.8}
$$

that is

$$
\dot{x} = Ax + B_1 w + B_2 u \tag{3.9}
$$
$$
z = C_1 x + D_{12} u \tag{3.10}
$$
$$
y = C_2 x + D_{21} w \tag{3.11}
$$

where x is the n-dimensional state vector, u is an m-vector of control inputs, w is an l-vector of external disturbance inputs, z is a p-vector of objectives, and y is a q-vector of measurements. We wish to find the proper, real-rational controller K, which stabilizes G internally, and minimizes

$$
\|T_{zw}(s)\|_2 \triangleq \sqrt{\frac{1}{2\pi} \int_{-\infty}^{\infty} Trace\left\{ T_{zw}^*(i\omega) T_{zw}(i\omega) \right\} d\omega},
$$

which is referred to as the $\mathcal{H}_2$-norm of T_{zw}. The transfer function T_{zw}, which is the transfer function from the external disturbances to the objectives, is given by

$$T_{zw}(s) = G_{11}(s) + G_{12}(s) K(s) (I - G_{22}(s) K(s))^{-1} G_{21}(s). \qquad (3.12)$$

The problem has a unique solution provided that

1. (A, B_2) is stabilizable and (C_2, A) is detectable;

2. D_{12} and D_{21} have full rank;

3. $\begin{bmatrix} A - i\omega I & B_2 \\ C_1 & D_{12} \end{bmatrix}$ has full column rank for all ω;

4. $\begin{bmatrix} A - i\omega I & B_1 \\ C_2 & D_{21} \end{bmatrix}$ has full row rank for all ω.

The optimal controller is given by

$$K_{opt}(s) = \left[\begin{array}{c|c} A_2 & -L_2 \\ \hline F_2 & 0 \end{array} \right]$$

where

$$\begin{array}{rcl} F_2 &=& -(B_2^* X_2 + D_{12}^* C_1) \\ L_2 &=& -(Y_2 C_2^* + B_1 D_{21}^*) \\ A_2 &=& A + B_2 F_2 + L_2 C_2 \end{array}$$

and X_2 and Y_2 uniquely solve the algebraic Riccati equations

$$(A - B_2 D_{12}^* C_1)^* X_2 + X_2 (A - B_2 D_{12}^* C_1)$$
$$- X_2 B_2 B_2^* X_2 + C_1^* C_1 - C_1^* D_{12} D_{12}^* C_1 = 0, \text{ and}$$

$$(A - B_1 D_{21}^* C_2)^* Y_2 + Y_2 (A^* - C_2^* D_{21} B_1^*)$$
$$- Y_2 C_2^* C_2 Y_2 + B_1 B_1^* - B_1 D_{21}^* D_{21} B_1^* = 0,$$

respectively. Note that the LQG control design of the previous section fits into this framework by setting

$$z = \begin{bmatrix} Q^{\frac{1}{2}} & 0 \\ 0 & R^{\frac{1}{2}} \end{bmatrix} \begin{bmatrix} x \\ u \end{bmatrix}$$

$$\begin{bmatrix} w_d \\ w_n \end{bmatrix} = \begin{bmatrix} W^{\frac{1}{2}} & 0 \\ 0 & V^{\frac{1}{2}} \end{bmatrix} w$$

where w is a white noise process of unit intensity. For this reason, the terms LQG and $\mathcal{H}_2$ control are used interchangeably in the literature.

3.1.4 $\mathcal{H}_\infty$ Control

Again, we consider the standard system (3.7) with the state-space realization (3.8) (sketched in Figure 3.2). The sub-optimal $\mathcal{H}_\infty$ control problem is to find all proper, real-rational controllers K, which stabilize (3.8) internally, and satisfy

$$\|T_{zw}(s)\|_\infty \triangleq \sup_\omega \sigma_{\max}(T_{zw}(j\omega)) < \gamma \tag{3.13}$$

for some prescribed γ, where $\sigma_{\max}(\cdot)$ denotes the maximum singular value of its argument. $\|T_{zw}(s)\|_\infty$ is referred to as the $\mathcal{H}_\infty$-norm of T_{zw}, and T_{zw} is given in (3.12). There is an optimal γ, denoted γ_0, which is simply the infimum of all γ for which the sub-optimal problem is solvable. Unlike the $\mathcal{H}_2$ control case, where we were able to explicitly write the unique optimal controller, the optimal γ must in this case be found iteratively. Returning to the sub-optimal problem, it is solvable provided Assumptions 1-4 in the previous section hold, and in addition, the following assumptions hold

5. X_∞ solves the algebraic Riccati equation

$$A^T X_\infty + X_\infty A - X_\infty \left(B_2 B_2^T - \frac{1}{\gamma^2} B_1 B_1^T \right) X_\infty + C_1^T C_1 = 0$$

and $X_\infty \geq 0$.

6. Y_∞ solves the algebraic Riccati equation

$$A Y_\infty + Y_\infty A^T - Y_\infty \left(C_2^T C_2 - \frac{1}{\gamma^2} C_1^T C_1 \right) Y_\infty + B_1 B_1^T = 0$$

and $Y_\infty \geq 0$.

7. $\rho(X_\infty Y_\infty) < \gamma^2$ (ρ denotes spectral radius).

Under these conditions, one admissible controller is given by

$$K_{sub}(s) = \left[\begin{array}{c|c} A_\infty & -Z_\infty L_\infty \\ \hline F_\infty & 0 \end{array} \right] \tag{3.14}$$

where

$$A_\infty = A + \frac{1}{\gamma^2} B_1 B_1^T X_\infty + B_2 F_\infty + Z_\infty L_\infty C_2$$

$$F_\infty = -B_2^T X_\infty, \ L_\infty = -Y_\infty C_2^T, \ Z_\infty = \left(I - \frac{1}{\gamma^2} Y_\infty X_\infty \right)^{-1}.$$

Note that in the limit $\gamma \to \infty$, the controller (3.14) approaches the optimal $\mathcal{H}_2$ controller of the previous section.

3.2   Nonlinear Systems

3.2.1   Stability in the Sense of Lyapunov

Consider the system

$$\dot{x} = f(t, x). \tag{3.15}$$

The following definitions are taken from [86, 89, 90].

Definition 3.1 *A continuous function $\alpha : \mathbb{R}_+ \to \mathbb{R}_+$ is said to be a class $\mathcal{K}$ function if it is strictly increasing and $\alpha(0) = 0$. If, in addition, $\alpha(r) \to \infty$ as $r \to \infty$ it is said to be a class $\mathcal{K}_\infty$ function. A continuous function $\beta : \mathbb{R}_+ \times \mathbb{R}_+ \to \mathbb{R}_+$ is said to be a class $\mathcal{KL}$ function if, for each fixed s, the mapping $\beta(r, s)$ is a class $\mathcal{K}$ function with respect to r and, for each fixed r, the mapping $\beta(r, s)$ is decreasing with respect to s and $\beta(r, s) \to 0$ as $s \to \infty$.*

Definition 3.2 *The solution $x(t; x_0, t_0)$ of (3.15) is uniformly globally bounded (UGB) if for each $x_0 \in \mathbb{R}^n$ there exists a constant b (independent of t_0), such that*

$$\|x(t)\| \leq b(x_0)$$

Definition 3.3 *The equilibrium point $x = 0$ of (3.15) is uniformly globally asymptotically stable (UGAS) if*

  1. *it is uniformly globally stable (UGS), that is, there exists $\gamma \in \mathcal{K}_\infty$ such that, for each $(t_0, x_0) \in \mathbb{R}_+ \times \mathbb{R}^n$ and all $t \geq t_0$, we have*

$$\|x(t)\| \leq \gamma\left(\|x_0\|\right);$$

  2. *for each pair of strictly positive real numbers ϵ and r, there exists a positive real number T such that*

$$\|x(t)\| \leq \epsilon, \quad \forall t \geq t_0 + T(\epsilon, r), \ \forall \|x(t_0)\| < r \tag{3.16}$$

Definition 3.4 *The equilibrium point $x = 0$ of (3.15) is uniformly locally exponentially stable (ULES) if there exist positive constants k, γ and c such that*

$$\|x(t)\| \leq k\|x(t_0)\| e^{-\gamma(t-t_0)}, \quad \forall t \geq t_0, \ \forall \|x(t_0)\| < c \tag{3.17}$$

If (3.17) is satisfied for any initial state $x(t_0)$, then the equilibrium point is uniformly globally exponentially stable (UGES).

Definition 3.5 *The system $\dot{x} = f(t, x, u)$ is input-to-state stable (ISS) if there exist a class $\mathcal{KL}$ function β, and a class $\mathcal{K}$ function γ, such that for any initial state $x(t_0)$, and any bounded input $u(t)$, the solution $x(t)$ exists and satisfies*

$$\|x(t)\| \leq \beta\left(\|x(t_0)\|, t - t_0\right) + \gamma\left(\sup_{t_0 \leq \tau \leq t} u(\tau)\right).$$

Definition 3.6 *The system $\dot{x} = f(t, x, u)$ with output $y = h(x)$ is input-output-to-state stable (IOSS) if there exist a class $\mathcal{KL}$ function β, and class $\mathcal{K}$ functions γ_1 and γ_2, such that for any initial state $x(t_0)$, and any bounded input $u(t)$, the solution $x(t)$ exists and satisfies*

$$\|x(t)\| \leq \max\left\{\beta\left(\|x(t_0)\|, t - t_0\right), \gamma_1\left(\sup_{t_0 \leq \tau \leq t} u(\tau)\right), \gamma_2\left(\sup_{t_0 \leq \tau \leq t} y(\tau)\right)\right\}.$$

Stability of solutions of system (3.15) is analyzed using Lyapunov stability theory [86, Theorem 3.8, Corollary 3.3, and Corollary 3.4].

Theorem 3.1 *Let $x = 0$ be an equilibrium point for the system $\dot{x} = f(t, x)$ and $D \subset \mathbb{R}^n$ be a domain containing $x = 0$. Let $V : [0, \infty) \times D \to \mathbb{R}$ be a continuously differentiable function such that*

$$W_1(x) \leq V(t, x) \leq W_2(x)$$
$$\frac{\partial V}{\partial t} + \frac{\partial V}{\partial x} f(t, x) \leq -W_3(x)$$

for all $t \geq 0$ and for all $x \in D$ where $W_1(x)$, $W_2(x)$, and $W_3(x)$ are continuous positive definite functions on D. Then , $x = 0$ is uniformly asymptotically stable. Moreover, if

$$W_1(x) \geq k_1 \|x\|^c, \ W_2(x) \leq k_2 \|x\|^c, \ W_3(x) \geq k_3 \|x\|^c$$

for some positive constants k_1, k_2, k_3, and c, then $x = 0$ is exponentially stable. If $D = \mathbb{R}^n$, then $x = 0$ is globally uniformly asymptotically stable (respectively globally exponentially stable).

The function V in Theorem 3.1 is called a *Lyapunov function*, and is in general not easy to find. A systematic method that in many cases simplifies the task of finding a Lyapunov function, called *backstepping*, is now presented in its simplest form (integrator backstepping).

3.2.2 Integrator Backstepping

Consider the system [90]

$$\dot{x} = f(x) + g(x)\xi, \ f(0) = 0 \tag{3.18}$$
$$\dot{\xi} = u, \tag{3.19}$$

where $x \in \mathbb{R}^n$ and $\xi, u \in \mathbb{R}$, and suppose there exist a continuously differentiable function $\alpha : \mathbb{R}^n \to \mathbb{R}$ with $\alpha(0) = 0$, and a smooth, positive definite, radially unbounded function $V : \mathbb{R}^n \to \mathbb{R}$, such that

$$\frac{\partial V}{\partial x}(x)[f(x) + g(x)\alpha(x)] \leq -W(x) \leq 0$$

where $W : \mathbb{R}^n \to \mathbb{R}$ is positive definite. In other words, if ξ is taken as the control input in (3.18), called the *virtual control*, then the control $\xi = \alpha(x)$ renders the equilibrium point $x = 0$ globally asymptotically stable. α is thus called a *stabilizing function*. We now define the *error variable* z to be the deviation of ξ from its desired value $\alpha(x)$, that is

$$z = \xi - \alpha(x).$$

Augmenting V with a quadratic term in z, we get the Lyapunov function candidate

$$V_a = V + \frac{1}{2}z^2$$

whose time derivative along solutions of system (3.18)–(3.19) is

$$
\begin{aligned}
\dot{V}_a &= \frac{\partial V}{\partial x}(x)[f(x) + g(x)\xi] + z\dot{z} \\
&= \frac{\partial V}{\partial x}(x)[f(x) + g(x)(z + \alpha(x))] + z\left(\dot{\xi} - \frac{\partial \alpha}{\partial x}(x)[f(x) + g(x)\xi]\right) \\
&= \frac{\partial V}{\partial x}(x)[f(x) + g(x)\alpha(x)] \\
&\quad + z\left[\frac{\partial V}{\partial x}(x)g(x) + u - \frac{\partial \alpha}{\partial x}(x)[f(x) + g(x)\xi]\right] \\
&\leq -W(x) + z\left[\frac{\partial V}{\partial x}(x)g(x) + u - \frac{\partial \alpha}{\partial x}(x)[f(x) + g(x)\xi]\right]. \tag{3.20}
\end{aligned}
$$

We are now in a position to select the control u in such a fashion that $\dot{V}_a \leq -W_a(x, z)$, with $W_a(x, z)$ positive definite. The most straight forward choice is simply to cancel all terms in the bracket in (3.20), and add a term that ensures positive definiteness of W_a. So, defining the control

$$u(x) = -c(\xi - \alpha(x)) - \frac{\partial V}{\partial x}(x)g(x) + \frac{\partial \alpha}{\partial x}(x)[f(x) + g(x)\xi] \tag{3.21}$$

we get $W_a(x, z) = W(x) + cz^2$, which proves that the control (3.21) renders $(x, z) = (0, 0)$ globally asymptotically stable, which in turn implies that $(x, \xi) = (0, 0)$ is globally asymptotically stable. If system (3.18)–(3.19) were augmented with more integrators, the backstepping procedure demonstrated above may be applied repeatedly, defining new virtual controls, stabilizing functions and error variables at each step. Such recursive designs are treated in detail in [90].

Chapter 4

Stabilization

The steady-state solutions, or equilibrium velocity profiles, obtained for the prototype flows of Section 2.6, are parabolic in shape in the streamwise direction, and zero in the other directions. Thus, the flows consist of parallel layers of fluid moving in a very regular and deterministic way. These are examples of so-called *laminar* flows. For wall-bounded laminar flows, wall friction, or *drag*, is favorably low, and these flows are therefore target flows in drag reduction applications. Unfortunately, they are rarely stable. In fact, stability is assured at small Reynolds number, only. An unstable flow is characterized by the fact that small perturbations from the equilibrium velocity profile will grow, and eventually cause the flow to transition to *turbulent* flow. Turbulent flow is characterized by small scale, apparently stochastic, velocity components, which lead to substantially higher drag than what is present in laminar flow. Being able to relaminarize a turbulent flow is therefore of great importance, and can be achieved in the prototype flows studied here, by stabilizing the parabolic equilibrium profile using boundary control. Boundary control implies specifying the flow field dynamically on the boundary of the domain, in this case on the channel or pipe walls, possibly based on values of flow variables taken at the boundary (boundary feedback). In this work, we assume that there exist sensors that provide distributed flow information at the wall, and actuators that can set prescribed distributed velocities. Chapter 6 reviews a selection of sensors and actuators that accomplish this task.

The problem of stabilizing the parabolic equilibrium profile of the channel flow has been attacked from several different angles by a number of authors. Approaches range from discretizing the linearized Navier-Stokes equation and using the tools available for stabilizing finite-dimensional linear time invariant systems, to Lyapunov stability analysis of the full, nonlinear Navier-Stokes equation. In the following sections, these efforts are summarized. Recent results on stabilization of pipe flows and cylinder flows are also presented.

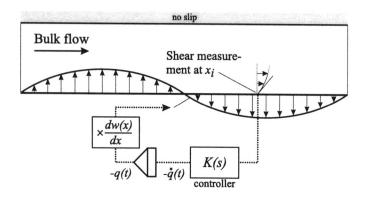

Figure 4.1: Control system configuration for controlling 2D channel flow by wall transpiration at the lower wall [78].

4.1 Linearization and Reduced Order Methods

This section summarizes the efforts on stabilization of the discretized, linearized Navier-Stokes equations. Bringing the linearized Navier-Stokes equations into the form of a linear time invariant system constitute a major part of the work involved in these methods. Once the state-space model is constructed, any tool from linear control theory can be applied in a fairly straight forward manner.

4.1.1 2D Channel Flow

In a series of papers [42, 43, 78, 79], stabilization of a reduced order model of 2D channel flow by classical and optimal control techniques is considered. The shear $(\partial U/\partial y)$ at a single point on the lower wall is taken as measurement, and the rate of change in the intensity of fluid transpiration on the lower wall is the control input. The actuator applies blowing and suction of fluid distributed as a prescribed function of x along the lower wall. The control system setup is shown schematically in Figure 4.1.

Reduced Order Model

The point of departure for obtaining a reduced order model in state space form is the linearized Navier-Stokes equation for 2D channel flow, (2.76) and (2.77)–(2.78). Due to (2.76) there exists a single valued function $\psi(x, y, t)$ such that

$$u(x, y, t) = \frac{\partial \psi}{\partial y}(x, y, t) \tag{4.1}$$

$$v(x, y, t) = -\frac{\partial \psi}{\partial x}(x, y, t). \tag{4.2}$$

The function ψ is unique up to a constant, and is called the *stream function* due to the fact that contours of constant ψ define streamlines of the flow. Inserting (4.1)–(4.2) into (2.77)–(2.78) yields

$$\frac{\partial^2 \psi}{\partial t \partial y} + \tilde{U}\frac{\partial^2 \psi}{\partial y \partial x} - \frac{\partial \tilde{U}}{\partial y}\frac{\partial \psi}{\partial x} + \frac{\partial p}{\partial x} - \frac{1}{Re}\left(\frac{\partial^3 \psi}{\partial y \partial x^2} + \frac{\partial^3 \psi}{\partial y^3}\right) = 0 \quad (4.3)$$

$$-\frac{\partial^2 \psi}{\partial t \partial x} - \tilde{U}\frac{\partial^2 \psi}{\partial x^2} + \frac{\partial p}{\partial y} + \frac{1}{Re}\left(\frac{\partial^3 \psi}{\partial x^3} + \frac{\partial^3 \psi}{\partial x \partial y^2}\right) = 0. \quad (4.4)$$

Taking the partial derivative of (4.3) with respect to y and subtracting the partial derivative of (4.4) with respect to x yield

$$\frac{\partial^3 \psi}{\partial t \partial y^2} + \frac{\partial^3 \psi}{\partial t \partial x^2} + \tilde{U}\frac{\partial^3 \psi}{\partial x \partial y^2} - \frac{\partial^2 \tilde{U}}{\partial y^2}\frac{\partial \psi}{\partial x} + \tilde{U}\frac{\partial^3 \psi}{\partial x^3}$$
$$- \frac{1}{Re}\left(\frac{\partial^4 \psi}{\partial x^4} + 2\frac{\partial^4 \psi}{\partial x^2 \partial y^2} + \frac{\partial^4 \psi}{\partial y^4}\right) = 0. \quad (4.5)$$

Suppose boundary control is applied by imposing a boundary condition on the lower wall of the form

$$\psi\left(x, y = -1, t\right) = q(t)w(x)f(y = -1) \quad (4.6)$$

while keeping no-slip boundary conditions on the upper wall, that is

$$\frac{\partial \psi}{\partial y}(x, y = 1, t) = \frac{\partial \psi}{\partial x}(x, y = 1, t) = 0.$$

Restricting the boundary control to blowing and suction of fluid through the wall at normal angle to the wall, that is $u(x, y = -1, t) = \frac{\partial \psi}{\partial y}(x, y = -1, t) = 0$, along with the no-slip condition on the upper wall, restrict the function $f(y)$ in the following manner

$$\frac{\partial f}{\partial y}(y = \pm 1) = 0 \quad (4.7)$$

$$f(y = 1) = 0. \quad (4.8)$$

As in [78], we will use

$$f(y) = \frac{1}{2}y^4 + \frac{1}{4}y^3 - y^2 - \frac{3}{4}y + 1$$

although there are many other choices satisfying (4.7)–(4.8). When implementing the boundary control law (4.6) in practice, the physical flow variables, U and V, must be set such that the resulting stream function satisfies (4.6). In terms of the physical flow variables U and V, the boundary control law is

$$U(x, y = \pm 1, t) = 0, \quad V(x, y = 1, t) = 0,$$

$$V(x, y = -1, t) = v(x, y = -1, t) =$$

$$-\frac{\partial \psi}{\partial x}(x, y = -1, t) = -q(t)\frac{\partial w}{\partial x}(x) f(y = -1).$$

In order to obtain homogeneous boundary conditions, we introduce the change of variables

$$\phi(x, y, t) \triangleq \psi(x, y, t) - q(t)w(x)f(y)$$

which substituted into (4.5) yields

$$\frac{\partial^3 \phi}{\partial t \partial y^2} + \frac{dq}{dt} w \frac{d^2 f}{dy^2} + \frac{\partial^3 \phi}{\partial t \partial x^2} + \frac{dq}{dt}\frac{d^2 w}{dx^2}f + \tilde{U}\left(\frac{\partial^3 \phi}{\partial x \partial y^2} + q\frac{dw}{dx}\frac{d^2 f}{dy^2}\right)$$

$$-\frac{d^2 \tilde{U}}{dy^2}\left(\frac{\partial \phi}{\partial x} + q\frac{dw}{dx}f\right) + \tilde{U}\left(\frac{\partial^3 \phi}{\partial x^3} + q\frac{d^3 w}{dx^3}f\right) - \frac{1}{Re}\left(\left(\frac{\partial^4 \phi}{\partial x^4} + q\frac{d^4 w}{dx^4}f\right)\right.$$

$$\left. +2\left(\frac{\partial^4 \phi}{\partial x^2 \partial y^2} + q\frac{d^2 w}{dx^2}\frac{d^2 f}{dy^2}\right) + \left(\frac{\partial^4 \phi}{\partial y^4} + qw\frac{d^4 f}{dy^4}\right)\right) = 0. \quad (4.9)$$

The boundary conditions in terms of ϕ are

$$\phi(x, y = \pm 1, t) = 0$$

$$\frac{\partial \phi}{\partial y}(x, y = \pm 1, t) = 0.$$

The streamwise component of shear at a single point at the lower wall is used as a measurement, that is

$$z(t) = \frac{\partial u}{\partial y}(x_i, y = -1, t).$$

In terms of ϕ, we have

$$z(t) = \frac{\partial^2 \psi}{\partial y^2}(x_i, y = -1, t) = \frac{\partial^2 \phi}{\partial y^2}(x_i, y = -1, t) + q(t)w(x_i)\frac{d^2 f}{dy^2}(y = -1).$$

$$(4.10)$$

A standard Fourier-Galerkin procedure is used to discretize the partial differential equation in the streamwise direction, and a Chebyshev collocation method is used in the wall-normal direction. Starting with the streamwise direction, we set

$$\phi(x, y, t) = \sum_{n=-N}^{N} a_n(y, t)P_n(x) \qquad (4.11)$$

$$w(x) = \sum_{n=-N}^{N} w_n P_n(x) \qquad (4.12)$$

where

$$P_n(x) = e^{in\frac{2\pi}{L}x}. \tag{4.13}$$

The partial derivatives of ϕ and w with respect to x up to order four are

$$\frac{\partial\phi}{\partial x}(x,y,t) = \sum_{n=-N}^{N}\alpha_n i a_n(y,t)P_n(x) \quad \frac{dw}{dx}(x) = \sum_{n=-N}^{N}\alpha_n i w_n P_n(x) \tag{4.14}$$

$$\frac{\partial^2\phi}{\partial x^2}(x,y,t) = -\sum_{n=-N}^{N}\alpha_n^2 a_n(y,t)P_n(x) \quad \frac{d^2 w}{dx^2}(x) = -\sum_{n=-N}^{N}\alpha_n^2 w_n P_n(x) \tag{4.15}$$

$$\frac{\partial^3\phi}{\partial x^3}(x,y,t) = -\sum_{n=-N}^{N}\alpha_n^3 i a_n(y,t)P_n(x) \quad \frac{d^3 w}{dx^3}(x) = -\sum_{n=-N}^{N}\alpha_n^3 i w_n P_n(x) \tag{4.16}$$

$$\frac{\partial^4\phi}{\partial x^4}(x,y,t) = \sum_{n=-N}^{N}\alpha_n^4 a_n(y,t)P_n(x) \quad \frac{d^4 w}{dx^4}(x) = \sum_{n=-N}^{N}\alpha_n^4 w_n P_n(x) \tag{4.17}$$

where $\alpha_n = 2\pi n/L$. The Gelerkin method provides the set of equations

$$\int_0^L \left\{ \frac{\partial^3\phi}{\partial t\partial y^2} + \frac{dq}{dt}w\frac{d^2 f}{dy^2} + \frac{\partial^3\phi}{\partial t\partial x^2} + \frac{dq}{dt}\frac{d^2 w}{dx^2}f + \tilde{U}\left(\frac{\partial^3\phi}{\partial x\partial y^2} + q\frac{dw}{dx}\frac{d^2 f}{dy^2}\right)\right.$$

$$- \frac{\partial^2\tilde{U}}{\partial y^2}\left(\frac{\partial\phi}{\partial x} + q\frac{dw}{dx}f\right) + \tilde{U}\left(\frac{\partial^3\phi}{\partial x^3} + q\frac{d^3 w}{dx^3}f\right) - \frac{1}{Re}\left(\frac{\partial^4\phi}{\partial x^4} + q\frac{d^4 w}{dx^4}f\right)$$

$$\left. +2\left(\frac{\partial^4\phi}{\partial x^2\partial y^2} + q\frac{d^2 w}{dx^2}\frac{d^2 f}{dy^2}\right) + \frac{\partial^4\phi}{\partial y^4} + qw\frac{d^4 f}{dy^4}\right\}\bar{P}_k(x)dx = 0. \tag{4.18}$$

Inserting (4.11)–(4.12) and (4.14)–(4.17) into (4.18) yields

$$\int_0^L \sum_{n=-N}^{N}\left\{\left[\frac{\partial^3 a_n}{\partial t\partial y^2} + \frac{dq}{dt}\frac{d^2 f}{dy^2}w_n - \alpha_n^2\frac{\partial a_n}{\partial t} - \frac{dq}{dt}f\alpha_n^2 w_n\right.\right.$$

$$+ \tilde{U}\left(\alpha_n i\frac{\partial^2 a_n}{\partial y^2} + q\frac{d^2 f}{dy^2}\alpha_n i w_n\right) - \frac{d^2\tilde{U}}{dy^2}(\alpha_n i a_n + q f\alpha_n i w_n)$$

$$- \tilde{U}(\alpha_n^3 i a_n + q f\alpha_n^3 i w_n) - \frac{1}{Re}[\alpha_n^4 a_n + q f\alpha_n^4 w_n$$

$$\left.\left. -2\left(\alpha_n^2\frac{\partial^2 a_n}{\partial y^2} + q\frac{d^2 f}{dy^2}\alpha_n^2 w_n\right) + \frac{\partial^4 a_n}{\partial y^4} + q\frac{d^4 f}{dy^4}w_n\right]\right]P_n\bar{P}_k\right\}dx = 0.$$

By the orthogonality of the P_n functions, that is

$$\int_0^L P_n(x)\bar{P}_k(x)\,dx = \int_0^L e^{in\frac{2\pi}{L}x}e^{-ik\frac{2\pi}{L}x}dx = \int_0^L e^{i(n-k)\frac{2\pi}{L}x}dx = \left\{\begin{array}{l} L, \; n = k \\ 0, \; n \neq k \end{array}\right.$$

we get

$$
\frac{\partial^3 a_k}{\partial t \partial y^2} + \frac{dq}{dt}\frac{d^2 f}{dy^2} w_k - \alpha_k^2 \frac{\partial a_k}{\partial t} - \frac{dq}{dt} f \alpha_k^2 w_k + \tilde{U}\left(\alpha_k i \frac{\partial^2 a_k}{\partial y^2} + q \frac{d^2 f}{dy^2}\alpha_k i w_k\right)
$$

$$
- \frac{\partial^2 \tilde{U}}{\partial y^2}\left(\alpha_k i a_k + q f \alpha_k i w_k\right) - \tilde{U}\left(\alpha_k^3 i a_k + q f \alpha_k^3 i w_k\right)
$$

$$
- \frac{1}{Re}\left[\alpha_k^4 a_k + q f \alpha_k^4 w_k - 2\left(\alpha_k^2 \frac{\partial^2 a_k}{\partial y^2} + q \frac{d^2 f}{dy^2}\alpha_k^2 w_k\right)\right.
$$

$$
\left. + \frac{\partial^4 a_k}{\partial y^4} + q \frac{d^4 f}{dy^4}w_k\right] = 0 \triangleq E, \text{ for } k = -N, ..., N. \quad (4.19)
$$

For the measurement (4.10), the Galerkin method yields

$$
z(t) = \sum_{k=-N}^{N} \frac{\partial^2 a_k}{\partial y^2}(y = -1, t) P_k(x_i) + \frac{5}{2}w(x_i)q. \quad (4.20)
$$

An interesting property of the set of equations (4.19) is that the equations are decoupled in terms of the wavenumber α_k. This fact lets us study each wavenumber individually. We now rewrite equation (4.19) in terms of a_k^R and a_k^I, where a_k^R and a_k^I are the real and imaginary parts of a_k, respectively. Since $real(a) = (a + \bar{a})/2$ and $imag(a) = (a - \bar{a})/2i$, the equations for a_k^R and a_k^I are obtained by

$$
\frac{1}{2}\left(E + \bar{E}\right) = 0
$$

and

$$
\frac{1}{2i}\left(E - \bar{E}\right) = 0,
$$

respectively. We have that

$$
\bar{E} = \frac{\partial^3 \bar{a}_k}{\partial t \partial y^2} + \frac{dq}{dt}\frac{d^2 f}{dy^2}\bar{w}_k - \alpha^2 \frac{\partial \bar{a}_k}{\partial t} - \frac{dq}{dt} f \alpha_k^2 \bar{w}_k - \tilde{U}\left(\alpha_k i \frac{\partial^2 \bar{a}_k}{\partial y^2} + q \frac{d^2 f}{dy^2}\alpha_k i \bar{w}_k\right)
$$

$$
+ \frac{d^2 \tilde{U}}{dy^2}\left(\alpha_k i \bar{a}_k + q f \alpha_k i \bar{w}_k\right) + \tilde{U}\left(\alpha_k^3 i \bar{a}_k + q f \alpha_k^3 i \bar{w}_k\right) - \frac{1}{Re}\left[\alpha_k^4 \bar{a}_k + q f \alpha_k^4 \bar{w}_k\right.
$$

$$
\left. - 2\left(\alpha_k^2 \frac{\partial^2 \bar{a}_k}{\partial y^2} + q \frac{d^2 f}{dy^2}\alpha_k^2 \bar{w}_k\right) + \frac{\partial^4 \bar{a}_k}{\partial y^4} + q \frac{d^4 f}{dy^4}\bar{w}_k\right]
$$

so that

$$
\frac{1}{2}\left(E + \bar{E}\right) = \frac{\partial^3 a_k^R}{\partial t \partial y^2} + \frac{dq}{dt}\frac{d^2 f}{dy^2}w_k^R - \alpha_k^2 \frac{\partial a_k^R}{\partial t} - \frac{dq}{dt}f\alpha_k^2 w_k^R
$$
$$
- \tilde{U}\left(\alpha_k \frac{\partial^2 a_k^I}{\partial y^2} + q\frac{d^2 f}{dy^2}\alpha_k w_k^I\right) + \frac{d^2\tilde{U}}{dy^2}\left(\alpha_k a_k^I + qf\alpha_k w_k^I\right)
$$
$$
+ \tilde{U}\left(\alpha_k^3 a_k^I + qf\alpha_k^3 w_k^I\right) - \frac{1}{Re}\Big[\alpha_k^4 a_k^R + qf\alpha_k^4 w_k^R
$$
$$
-2\left(\alpha_k^2\frac{\partial^2 a_k^R}{\partial y^2} + q\frac{d^2 f}{dy^2}\alpha_k^2 w_k^R\right) + \frac{\partial^4 a_k^R}{\partial y^4} + q\frac{d^4 f}{dy^4}w_k^R\Big] = 0 \quad (4.21)
$$

and

$$
\frac{1}{2i}\left(E - \bar{E}\right) = \frac{\partial^3 a_k^I}{\partial t \partial y^2} + \frac{dq}{dt}\frac{d^2 f}{dy^2}w_k^I - \alpha_k^2 \frac{\partial a_k^I}{\partial t} - \frac{dq}{dt}f\alpha_k^2 w_k^I
$$
$$
+ \tilde{U}\left(\alpha_k \frac{\partial^2 a_k^R}{\partial y^2} + q\frac{d^2 f}{dy^2}\alpha_k w_k^R\right) - \frac{d^2\tilde{U}}{dy^2}\left(\alpha_k a_k^R + qf\alpha_k w_k^R\right)
$$
$$
- \tilde{U}\left(\alpha_k^3 a_k^R + qf\alpha_k^3 w_k^R\right) - \frac{1}{Re}\Big[\alpha_k^4 a_k^I + qf\alpha_k^4 w_k^I
$$
$$
-2\left(\alpha_k^2\frac{\partial^2 a_k^I}{\partial y^2} + q\frac{d^2 f}{dy^2}\alpha_k^2 w_k^I\right) + \frac{\partial^4 a_k^I}{\partial y^4} + q\frac{d^4 f}{dy^4}w_k^I\Big] = 0. \quad (4.22)
$$

Next, we discretize the equation in the y-direction. For this, we use the Chebyshev collocation method described in Section 2.7.3 on $N + 1$ Chebyshev-Gauss-Labotto points as defined in (2.89). Applying the differentiation matrix $\mathcal{D}_N$, treating f and $\tilde{U}$ and their derivatives as known functions, yields

$$
\mathcal{D}_N^2 \frac{d\mathbf{a}_k^R}{dt} - \alpha_k^2 \frac{d\mathbf{a}_k^R}{dt} = \frac{dq}{dt}\mathbf{f}_N^{(0)}\alpha_k^2 w_k^R - \frac{dq}{dt}\mathbf{f}_N^{(2)}w_k^R + \tilde{U}_N\left(\alpha_k \mathcal{D}_N^2 \mathbf{a}_k^I + q\mathbf{f}_N^{(2)}\alpha_k w_k^I\right)
$$
$$
-\tilde{U}_N^{(2)}\left(\alpha_k \mathbf{a}_k^I + q\mathbf{f}_N^{(0)}\alpha_k w_k^I\right) - \tilde{U}_N\left(\alpha_k^3 \mathbf{a}_k^I + q\mathbf{f}_N^{(0)}\alpha_k^3 w_k^I\right) + \frac{1}{Re}\Big[\alpha_k^4 \mathbf{a}_k^R + q\mathbf{f}_N^{(0)}\alpha_k^4 w_k^R
$$
$$
-2\left(\alpha_k^2 \mathcal{D}_N^2 \mathbf{a}_k^R + q\mathbf{f}_N^{(2)}\alpha_k^2 w_k^R\right) + \mathcal{D}_N^4 \mathbf{a}_k^R + q\mathbf{f}_N^{(4)}w_k^R\Big] \quad (4.23)
$$

$$
\mathcal{D}_N^2 \frac{d\mathbf{a}_k^I}{dt} - \alpha_k^2 \frac{d\mathbf{a}_k^I}{dt} = \frac{dq}{dt}\mathbf{f}_N^{(0)}\alpha_k^2 w_k^I - \frac{dq}{dt}\mathbf{f}_N^{(2)}w_k^I - \tilde{U}_N\left(\alpha_k \mathcal{D}_N^2 \mathbf{a}_k^R + q\mathbf{f}_N^{(2)}\alpha_k w_k^R\right)
$$
$$
+\tilde{U}_N^{(2)}\left(\alpha_k \mathbf{a}_k^R + q\mathbf{f}_N^{(0)}\alpha_k w_k^R\right) + \tilde{U}_N\left(\alpha_k^3 \mathbf{a}_k^R + q\mathbf{f}_N^{(0)}\alpha_k^3 w_k^R\right) + \frac{1}{Re}\Big[\alpha_k^4 \mathbf{a}_k^I + q\mathbf{f}_N^{(0)}\alpha_k^4 w_k^I
$$
$$
-2\left(\alpha_k^2 \mathcal{D}_N^2 \mathbf{a}_k^I + q\mathbf{f}_N^{(2)}\alpha_k^2 w_k^I\right) + \mathcal{D}_N^4 \mathbf{a}_k^I + q\mathbf{f}_N^{(4)}w_k^I\Big] \quad (4.24)
$$

where $\mathbf{a}_k^R$ and $\mathbf{a}_k^I$ denote the vectors

$$
\mathbf{a}_k^R = \begin{bmatrix} a_k^R(y_0) & a_k^R(y_1) & \cdots & a_k^R(y_N) \end{bmatrix}^T
$$
$$
\mathbf{a}_k^I = \begin{bmatrix} a_k^I(y_0) & a_k^I(y_1) & \cdots & a_k^I(y_N) \end{bmatrix}^T
$$

and

$$
\tilde{U}_N^{(m)} = \begin{bmatrix} \frac{d^m \tilde{U}}{dy^m}(y_0) & 0 & 0 & 0 \\ 0 & \frac{d^m \tilde{U}}{dy^m}(y_1) & 0 & 0 \\ 0 & 0 & \ddots & 0 \\ 0 & 0 & 0 & \frac{d^m \tilde{U}}{dy^m}(y_N) \end{bmatrix}, \quad m = 0, 1, 2 \ldots \quad (4.25)
$$

$$
\mathbf{f}_N^{(m)} = \begin{bmatrix} \frac{d^m f}{dy^m}(y_0) & \frac{d^m f}{dy^m}(y_1) & \cdots & \frac{d^m f}{dy^m}(y_N) \end{bmatrix}^T, \quad m = 0, 1, 2, \ldots
$$

Above, (m) denotes the mth order derivative with respect to y. It remains to implement the boundary conditions. The Dirichlet boundary conditions are satisfied by simply setting $a_k^R(y_0) = a_k^I(y_0) = a_k^R(y_N) = a_k^I(y_N) = 0$, and omitting the differential equations for these variables. Satisfying the Neumann boundary conditions is more involved. We start by noticing that the Neumann boundary conditions imply that

$$
\sum_{j=0}^N d_{0j} a_k(y_j) = 0, \text{ and } \sum_{j=0}^N d_{Nj} a_k(y_j) = 0 \quad (4.26)
$$

where the superscripts I and R are omitted since the following derivation is valid for either one. The d_{jk} constants in (4.26) are the elements of $\mathcal{D}_N$ as defined in (2.90). From (4.26), we can solve for $a_k(y_1)$ and $a_k(y_{N-1})$ to obtain

$$
a_k(y_1) = \mathbf{l}_1^T \mathbf{a}
$$
$$
a_k(y_{N-1}) = \mathbf{l}_2^T \mathbf{a}
$$

where we have defined

$$
\begin{bmatrix} \mathbf{l}_1^T \\ \mathbf{l}_2^T \end{bmatrix} = - \begin{bmatrix} d_{01} & d_{0(N-1)} \\ d_{N1} & d_{N(N-1)} \end{bmatrix}^{-1} \begin{bmatrix} d_{02} & d_{03} & \cdots & d_{0(N-2)} \\ d_{N2} & d_{N3} & \cdots & d_{N(N-2)} \end{bmatrix}
$$

and

$$
\mathbf{a} = \begin{bmatrix} a_k(y_2) & a_k(y_3) & \cdots & a_k(y_{N-2}) \end{bmatrix}^T.
$$

Thus, we have that

$$
\mathbf{a}_k = \begin{bmatrix} \mathbf{0}^T \\ \mathbf{l}_1^T \\ \mathcal{I}_{N-3} \\ \mathbf{l}_2^T \\ \mathbf{0}^T \end{bmatrix} \mathbf{a} \triangleq \mathcal{I}_a \mathbf{a}. \quad (4.27)
$$

Inserting (4.27) into (4.23)–(4.24), and keeping in mind that the equations for $a_k(y_0)$, $a_k(y_1)$, $a_k(y_{N-1})$, and $a_k(y_N)$ should be omitted since these variables are determined by the boundary conditions, we obtain

$$
\begin{aligned}
R\left(\mathcal{D}_N^2 - \alpha_k^2 \mathcal{I}\right) \mathcal{I}_a \dot{\mathbf{a}}^R &= \frac{1}{Re} R\left(\mathcal{D}_N^4 \mathcal{I}_a - 2\alpha_k^2 \mathcal{D}_N^2 \mathcal{I}_a + \alpha_k^4 \mathcal{I}_a\right) \mathbf{a}^R \\
&\quad + R\left(\alpha_k \tilde{U}_N \mathcal{D}_N^2 \mathcal{I}_a - \left(\alpha_k \tilde{U}_N^{(2)} + \alpha_k^3 \tilde{U}_N\right) \mathcal{I}_a\right) \mathbf{a}^I \\
&\quad + w_k^I R\left(\alpha_k \tilde{U}_N \mathbf{f}_N^{(2)} - \left(\alpha_k \tilde{U}_N^{(2)} + \alpha_k^3 \tilde{U}_N\right) \mathbf{f}_N^{(0)}\right) q \\
&\quad + \frac{w_k^R}{Re} R\left(\alpha_k^4 \mathbf{f}_N^{(0)} - 2\alpha_k^2 \mathbf{f}_N^{(2)} + \mathbf{f}_N^{(4)}\right) q - w_k^R R\left(\mathbf{f}_N^{(2)} - \alpha_k^2 \mathbf{f}_N^{(0)}\right) \dot{q}
\end{aligned}
$$

$$
\begin{aligned}
R\left(\mathcal{D}_N^2 - \alpha_k^2 \mathcal{I}\right) \mathcal{I}_a \dot{\mathbf{a}}^I &= -R\left(\alpha_k \tilde{U}_N \mathcal{D}_N^2 \mathcal{I}_a - \left(\alpha_k \tilde{U}_N^{(2)} + \alpha_k^3 \tilde{U}_N\right) \mathcal{I}_a\right) \mathbf{a}^R \\
&\quad + \frac{1}{Re} R\left(\mathcal{D}_N^4 \mathcal{I}_a - 2\alpha_k^2 \mathcal{D}_N^2 \mathcal{I}_a + \alpha_k^4 \mathcal{I}_a\right) \mathbf{a}^I \\
&\quad - w_k^R R\left(\alpha_k \tilde{U}_N \mathbf{f}_N^{(2)} - \left(\alpha_k \tilde{U}_N^{(2)} + \alpha_k^3 \tilde{U}_N\right) \mathbf{f}_N^{(0)}\right) q \\
&\quad + \frac{w_k^I}{Re} R\left(\alpha_k^4 \mathbf{f}_N^{(0)} - 2\alpha_k^2 \mathbf{f}_N^{(2)} + \mathbf{f}_N^{(4)}\right) q - w_k^I R\left(\mathbf{f}_N^{(2)} - \alpha_k^2 \mathbf{f}_N^{(0)}\right) \dot{q}
\end{aligned}
$$

where R is a matrix that selects appropriate rows, defined as

$$
R = \left[\begin{array}{ccc} \mathbf{0}_{(N-3)\times 2} & \mathcal{I}_{N-3} & \mathbf{0}_{(N-3)\times 2} \end{array} \right].
$$

Assuming the measurement is real, we redefine z as the real part of (4.20), so we get

$$
z(t) = \sum_{k=-M}^{M} \left\{ \cos\left(\alpha_k x_i\right) \mathbf{s}_1^T \mathcal{D}_N^2 \mathcal{I}_a \mathbf{a}^R - \sin\left(\alpha_k x_i\right) \mathbf{s}_1^T \mathcal{D}_N^2 \mathcal{I}_a \mathbf{a}^I \right\} + \frac{5}{2} w(x_i) q
\tag{4.28}
$$

where $\mathbf{s}_j$ is a vector of compatible dimension with a 1 at the jth position being the only nonzero entry. It is clear that the measurement (4.28) is not decoupled in terms of the wavenumber α_k, but if we choose w appropriately, we may omit all terms but one in (4.28). The reason for this will be discussed in the next section. Finally we get the system in state space form

$$
\begin{aligned}
\dot{\mathbf{x}} &= A\mathbf{x} + Bu \tag{4.29} \\
z &= C\mathbf{x} + Du \tag{4.30}
\end{aligned}
$$

where

$$
\mathbf{x} = \left[\begin{array}{cccccc} a_k^R(y_2) & \cdots & a_k^R(y_{N-2}) & a_k^I(y_2) & \cdots & a_k^I(y_{N-2}) & q \end{array} \right]^T, \quad u = \dot{q}
$$

$$
A = \begin{bmatrix} M^{-1}A_R^R & M^{-1}A_R^I & M^{-1}\left(w_k^I \mathbf{q}_1 + w_k^R \mathbf{q}_2\right) \\ M^{-1}A_I^R & M^{-1}A_I^I & M^{-1}\left(-w_k^R \mathbf{q}_1 + w_k^I \mathbf{q}_2\right) \\ 0 & 0 & 0 \end{bmatrix}, \quad B = \begin{bmatrix} -w_k^R M^{-1}\mathbf{b} \\ -w_k^I M^{-1}\mathbf{b} \\ 1 \end{bmatrix}
$$

$$C = \left[\ \cos\left(\alpha_k x_i\right) \mathbf{s}_N^T \mathcal{D}_N^2 \mathcal{I}_a \quad -\sin\left(\alpha_k x_i\right) \mathbf{s}_N^T \mathcal{D}_N^2 \mathcal{I}_a \quad \tfrac{5}{2} w(x_i) \ \right], \ D = 0$$

$$
\begin{aligned}
M &= R\left(\mathcal{D}_N^2 - \alpha_k^2 \mathcal{I}\right)\mathcal{I}_a \\
A_R^R &= A_I^I = \frac{1}{Re} R\left(\mathcal{D}_N^4 \mathcal{I}_a - 2\alpha_k^2 \mathcal{D}_N^2 \mathcal{I}_a + \alpha_k^4 \mathcal{I}_a\right) \\
A_R^I &= -A_I^R = R\left(\alpha_k \tilde{U}_N \mathcal{D}_N^2 \mathcal{I}_a - \left(\alpha_k \tilde{U}_N^{(2)} + \alpha_k^3 \tilde{U}_N\right)\mathcal{I}_a\right)
\end{aligned}
$$

$$
\begin{aligned}
\mathbf{q}_1 &= R\left(\alpha_k \tilde{U}_N \mathbf{f}_N^{(2)} - \left(\alpha_k \tilde{U}_N^{(2)} + \alpha_k^3 \tilde{U}_N\right)\mathbf{f}_N^{(0)}\right) \\
\mathbf{q}_2 &= \frac{1}{Re} R\left(\alpha_k^4 \mathbf{f}_N^{(0)} - 2\alpha_k^2 \mathbf{f}_N^{(2)} + \mathbf{f}_N^{(4)}\right) \\
\mathbf{b} &= R\left(\mathbf{f}_N^{(2)} - \alpha_k^2 \mathbf{f}_N^{(0)}\right), \ R = \left[\ \mathbf{0}_{(N-3)\times 2} \quad \mathcal{I}_{N-3} \quad \mathbf{0}_{(N-3)\times 2} \ \right].
\end{aligned}
$$

Classical Control

In the previous section we developed a state space representation of the discretized 2D linearized channel flow equations. Apart from the measurement equation, the resulting system is decoupled in terms of the wave number. It turns out that the system has one complex conjugate pair of eigenvalues in the right half of the complex plane for $k = 2$. All other poles of the system are in the left half of the complex plane (except for the pole at the origin, due to the boundary control input configuration). Figure 4.2 shows poles and zeros for $k = 2$ at Reynolds number 10000 and $N = 150$ (i.e. 151 collocation points), and with the shear measurement taken at $x_i = \pi$, $w(x) = sin(\frac{2\pi k}{L}x)$, and $L = 4\pi$. With this choice of $w(x)$, (A, B) is rendered uncontrollable for all wave numbers with $k \neq 2$. Therefore, the fact that the measurement does not decouple in terms of the wavenumber does not present a problem, since the control will be unable to destabilize any pole of sub-systems with $k \neq 2$. Figure 4.3 shows the Nyquist plot. The Nyquist stability criterion indicates that when $-1/K \in (-105, 0)$ (approximately), no poles of the closed loop system lie in the right half plane, which corresponds to $K > 0.0095$. This result is confirmed by Figure 4.4, which shows the real component of the system's least stable eigenvalue as a function of feedback gain K.

LQG Control

Assuming that (4.29)–(4.30) are subjected to additive disturbances, w_d and w_n, that are uncorrelated Gaussian stochastic processes with zero means and covariances W, and V, respectively, it is straight forward to apply LQG control theory. For illustration, we select $W = I$ and $V = 1$, and $Q = I$ and $R = 1$, and construct the controller according to (3.6). Figure 4.5 shows the result in terms of the system poles for the open-loop system, the state feedback closed-loop

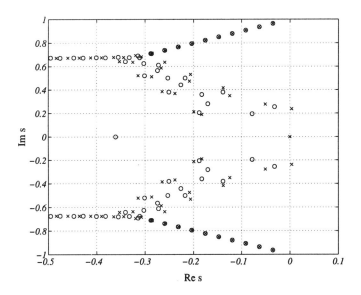

Figure 4.2: Poles ($\times$) and zeros ($\circ$) for the system [78].

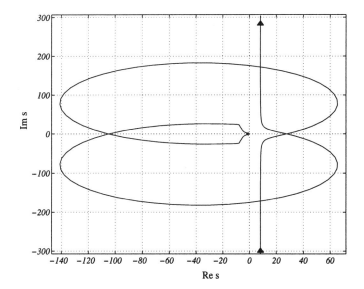

Figure 4.3: Nyquist plot.

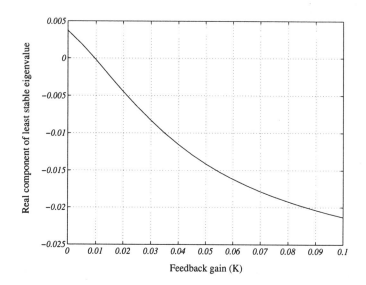

Figure 4.4: Real component of the least stable eigenvalue as a function of feedback gain K.

system, and the output feedback closed-loop system. Clearly, the open-loop unstable poles are moved into the left half of the complex plane by the control.

Implementation

Although the state space model (4.29)–(4.30) takes the form of a single-input-single-output (SISO) system, with input $u = \dot{q}(t)$ and output $z = \partial u/\partial y \, (x_i,$ $-1,\ t)$, the realization of the control system is distributed. This follows from the fact that the actuation is blowing and suction of fluid distributed along the wall as a *cosine* in x (dw/dx, where w is a sine in x), whose amplitude is altered with a speed given by the control input $u = \dot{q}(t)$. The control law is centralized since the control signal must reach all actuators on the wall. Centralized versus decentralized control schemes are discussed in more detail in Section 4.2.

4.1.2   3D Channel Flow

Reduced Order Model

In [24] linear control is applied to the 3D channel flow. In this case, shear measurements in two directions ($\partial U/\partial y$ and $\partial W/\partial y$) are taken at every point on both walls ($y = \pm 1$), and actuation is applied in the form of wall transpiration on both walls. The point of departure is the linearized Navier-Stokes equations

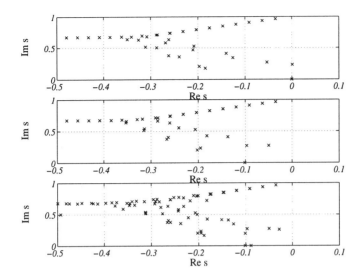

Figure 4.5: LQG synthesis: system poles for the open-loop system (top), the state feedback closed-loop system (middle), and the output feedback closed-loop system (bottom) [78].

for 3D channel flow (2.55)–(2.58), which may be rewritten in terms of the wall-normal velocity component v, and the wall-normal vorticity component, denoted ω (vorticity is defined as $\omega = curl(\mathbf{W})$). Taking the Laplacian of (2.57) yields

$$
\frac{\partial}{\partial t}\left(\frac{\partial^2 v}{\partial x^2} + \frac{\partial^2 v}{\partial y^2} + \frac{\partial^2 v}{\partial z^2}\right) + \tilde{U}\frac{\partial}{\partial x}\left(\frac{\partial^2 v}{\partial x^2} + \frac{\partial^2 v}{\partial y^2} + \frac{\partial^2 v}{\partial z^2}\right)
$$
$$
+ \frac{\partial^2 \tilde{U}}{\partial y^2}\frac{\partial v}{\partial x} + 2\frac{\partial \tilde{U}}{\partial y}\frac{\partial^2 v}{\partial x \partial y} = -\frac{\partial}{\partial y}\left(\frac{\partial^2 p}{\partial x^2} + \frac{\partial^2 p}{\partial y^2} + \frac{\partial^2 p}{\partial z^2}\right)
$$
$$
+ \frac{1}{Re}\left(\frac{\partial^4 v}{\partial x^4} + \frac{\partial^4 v}{\partial y^4} + \frac{\partial^4 v}{\partial z^4} + 2\frac{\partial^4 v}{\partial x^2 \partial y^2} + 2\frac{\partial^4 v}{\partial x^2 \partial z^2} + 2\frac{\partial^4 v}{\partial y^2 \partial z^2}\right). \quad (4.31)
$$

Taking the divergence of (2.56)–(2.58) yields

$$
\frac{\partial}{\partial t}\left(\frac{\partial u}{\partial x} + \frac{\partial v}{\partial y} + \frac{\partial w}{\partial z}\right) + \tilde{U}\frac{\partial}{\partial x}\left(\frac{\partial u}{\partial x} + \frac{\partial v}{\partial y} + \frac{\partial w}{\partial z}\right) + 2\frac{d\tilde{U}}{dy}\frac{\partial v}{\partial x} =
$$
$$
-\frac{\partial^2 p}{\partial x^2} - \frac{\partial^2 p}{\partial y^2} - \frac{\partial^2 p}{\partial z^2} + \frac{1}{Re}\left(\frac{\partial^2}{\partial x^2}\left(\frac{\partial u}{\partial x} + \frac{\partial v}{\partial y} + \frac{\partial w}{\partial z}\right)\right.
$$
$$
\left. + \frac{\partial^2}{\partial y^2}\left(\frac{\partial u}{\partial x} + \frac{\partial v}{\partial y} + \frac{\partial w}{\partial z}\right) + \frac{\partial^2}{\partial z^2}\left(\frac{\partial u}{\partial x} + \frac{\partial v}{\partial y} + \frac{\partial w}{\partial z}\right)\right)
$$

and using continuity (2.55), we obtain

$$2\frac{d\tilde{U}}{dy}\frac{\partial v}{\partial x} = -\frac{\partial^2 p}{\partial x^2} - \frac{\partial^2 p}{\partial y^2} - \frac{\partial^2 p}{\partial z^2}. \tag{4.32}$$

Inserting (4.32) into (4.31) yields the equation for the wall-normal velocity component, v, as

$$\frac{\partial}{\partial t}\left(\frac{\partial^2 v}{\partial x^2} + \frac{\partial^2 v}{\partial y^2} + \frac{\partial^2 v}{\partial z^2}\right) = -\tilde{U}\frac{\partial}{\partial x}\left(\frac{\partial^2 v}{\partial x^2} + \frac{\partial^2 v}{\partial y^2} + \frac{\partial^2 v}{\partial z^2}\right) + \frac{d^2\tilde{U}}{dy^2}\frac{\partial v}{\partial x}$$

$$+ \frac{1}{Re}\left(\frac{\partial^4 v}{\partial x^4} + \frac{\partial^4 v}{\partial y^4} + \frac{\partial^4 v}{\partial z^4} + 2\frac{\partial^4 v}{\partial x^2 \partial y^2} + 2\frac{\partial^4 v}{\partial x^2 \partial z^2} + 2\frac{\partial^4 v}{\partial y^2 \partial z^2}\right). \tag{4.33}$$

The wall-normal component of vorticity is $\omega = \partial u/\partial z - \partial w/\partial x$. Thus, taking the partial derivative of (2.56) with respect to z and subtracting the partial derivative of (2.58) with respect to x, yields

$$\frac{\partial \omega}{\partial t} = -\frac{d\tilde{U}}{dy}\frac{\partial v}{\partial z} - \tilde{U}\frac{\partial \omega}{\partial x} + \frac{1}{Re}\left(\frac{\partial^2 \omega}{\partial x^2} + \frac{\partial^2 \omega}{\partial y^2} + \frac{\partial^2 \omega}{\partial z^2}\right). \tag{4.34}$$

The control is applied as an unsteady boundary condition on the wall-normal velocity component v, and the no-slip condition in the x and z directions leads to homogeneous Dirichlet boundary conditions for ω, that is

$$\omega\left(x, y = \pm 1, z, t\right) = 0.$$

By continuity, $\partial v/\partial y = 0$ at the wall, so (4.33) is also subject to homogeneous Neumann boundary conditions. Expanding v and ω as

$$v(x, y, z, t) = \sum_{k_x, k_z} \hat{v}\left(k_x, y, k_z, t\right) e^{i\left(\frac{2\pi k_x}{L_x}x + \frac{2\pi k_z}{L_z}z\right)}$$

$$\omega(x, y, z, t) = \sum_{k_x, k_z} \hat{\omega}\left(k_x, y, k_z, t\right) e^{i\left(\frac{2\pi k_x}{L_x}x + \frac{2\pi k_z}{L_z}z\right)},$$

the Fourier-Galerkin method yields

$$\frac{\partial}{\partial t}\left(-\alpha_x^2\hat{v} + \frac{\partial^2 \hat{v}}{\partial y^2} - \alpha_z^2\hat{v}\right) = \tilde{U}\left(i\alpha_x^3\hat{v} - i\alpha_x\frac{\partial^2 \hat{v}}{\partial y^2} + i\alpha_x\alpha_z^2\hat{v}\right)$$

$$+ \frac{\partial^2 \tilde{U}}{\partial y^2}i\alpha_x\hat{v} + \frac{1}{Re}\left(\alpha_x^4\hat{v} + \frac{\partial^4 \hat{v}}{\partial y^4} + \alpha_z^4\hat{v} - 2\alpha_x^2\frac{\partial^2 \hat{v}}{\partial y^2} + 2\alpha_x^2\alpha_z^2\hat{v} - 2\alpha_z^2\frac{\partial^2 \hat{v}}{\partial y^2}\right)$$

$$\frac{\partial \hat{\omega}}{\partial t} = -\frac{\partial \tilde{U}}{\partial y}i\alpha_z\hat{v} - \tilde{U}i\alpha_x\hat{\omega} + \frac{1}{Re}\left(-\alpha_x^2\hat{\omega} + \frac{\partial^2 \hat{\omega}}{\partial y^2} - \alpha_z^2\hat{\omega}\right)$$

where

$$\alpha_x = \frac{2\pi k_x}{L_x}$$

$$\alpha_z = \frac{2\pi k_z}{L_x}.$$

As in Section 4.1.1 we discretize the wall-normal direction on $N+1$ Chebyshev-Gauss-Lobatto (2.89) points using the Chebyshev collocation method. Applying the differentiation matrix $\mathcal{D}_N$, treating $\tilde{U}$ and its derivatives as known functions, yields

$$\left(\mathcal{D}_N^2 - \left(\alpha_x^2 + \alpha_z^2\right)\mathcal{I}\right)\frac{d\hat{\mathbf{v}}_a}{dt} = \left(\tilde{U}_N^{(0)}\left(-i\alpha_x\mathcal{D}_N^2 + \left(i\alpha_x^3 + i\alpha_x\alpha_z^2\right)\mathcal{I}\right)\right.$$

$$\left.+\tilde{U}_N^{(2)}i\alpha_x + \frac{1}{Re}\left(\mathcal{D}_N^4 - 2\left(\alpha_x^2 + \alpha_z^2\right)\mathcal{D}_N^2 + \left(\alpha_x^2 + \alpha_z^2\right)^2\mathcal{I}\right)\right)\hat{\mathbf{v}}_a \quad (4.35)$$

$$\frac{d\hat{\omega}_a}{dt} = -\tilde{U}_N^{(1)}i\alpha_z\hat{\mathbf{v}}_a + \left(-\tilde{U}_N^{(0)}i\alpha_x + \frac{1}{Re}\left(\mathcal{D}_N^2 - \left(\alpha_x^2 + \alpha_z^2\right)\mathcal{I}\right)\right)\hat{\omega}_a \quad (4.36)$$

where

$$\hat{\mathbf{v}}_a = \begin{bmatrix} \hat{v}(y_0) & \hat{v}(y_1) & \cdots & \hat{v}(y_N) \end{bmatrix}^T$$

$$\hat{\omega}_a = \begin{bmatrix} \hat{\omega}(y_0) & \hat{\omega}(y_1) & \cdots & \hat{\omega}(y_N) \end{bmatrix}^T$$

and $\tilde{U}_N^{(m)}$ is defined as in (4.25). Since the time derivative of $\hat{v}(y_0)$ and $\hat{v}(y_N)$ occur in (4.35)–(4.36), we let the time derivative of $\hat{v}(y_0)$ and $\hat{v}(y_N)$ be the control input. The Neumann boundary conditions on v imply that

$$\sum_{j=0}^{N} d_{0j}\hat{v}(y_j) = 0, \text{ and } \sum_{j=0}^{N} d_{Nj}\hat{v}(y_j) = 0. \quad (4.37)$$

Solving (4.37) for $\hat{v}(y_1)$ and $\hat{v}(y_{N-1})$, yields

$$\hat{v}(y_1) = \mathbf{l}_1^T\hat{\mathbf{v}} \quad (4.38)$$

$$\hat{v}(y_{N-1}) = \mathbf{l}_2^T\hat{\mathbf{v}} \quad (4.39)$$

where we have defined

$$\begin{bmatrix} \mathbf{l}_1^T \\ \mathbf{l}_2^T \end{bmatrix} = -\begin{bmatrix} d_{01} & d_{0(N-1)} \\ d_{N1} & d_{N(N-1)} \end{bmatrix}^{-1}\begin{bmatrix} d_{00} & d_{02} & \cdots & d_{0(N-2)} & d_{0N} \\ d_{N0} & d_{N2} & \cdots & d_{N(N-2)} & d_{NN} \end{bmatrix}$$

$$\hat{\mathbf{v}} = \begin{bmatrix} \hat{v}(y_0) & \hat{v}(y_2) & \cdots & \hat{v}(y_{N-2}) & \hat{v}(y_N) \end{bmatrix}^T.$$

Thus, we get

$$\hat{\mathbf{v}}_a = \begin{bmatrix} \mathbf{s}_1^T \\ \mathbf{l}_1^T \\ \mathcal{I}_{N-3} \\ \mathbf{l}_2^T \\ \mathbf{s}_{N-3}^T \end{bmatrix}\hat{\mathbf{v}} \triangleq \mathcal{I}_v\hat{\mathbf{v}} \quad (4.40)$$

and, defining $\hat{\omega} \triangleq \begin{bmatrix} \hat{\omega}(y_1) & \hat{\omega}(y_2) & \cdots & \hat{\omega}(y_{N-1}) \end{bmatrix}^T$, the homogeneous Dirichlet boundary conditions on ω yields

$$\hat{\omega}_a = \begin{bmatrix} 0^T \\ \mathcal{I}_{N-1} \\ 0^T \end{bmatrix} \hat{\omega} \triangleq \mathcal{I}_\omega \hat{\omega}. \tag{4.41}$$

Inserting (4.40) and (4.41) into (4.35)–(4.36), omitting the equations for $\hat{v}(y_1)$, and $\hat{v}(y_{N-1})$ (since they are determined by the homogeneous Neumann boundary conditions), and omitting the equations for $\hat{\omega}(y_0)$ and $\hat{\omega}(y_N)$ (since they are zero due to the homogeneous Dirichlet boundary conditions), we get

$$\frac{d\hat{v}}{dt} = M_1^{-1} R_1 \left(\tilde{U}_N^{(0)} \left(-i\alpha_x \mathcal{D}_N^2 + i\alpha_x \alpha \mathcal{I} \right) + \tilde{U}_N^{(2)} i\alpha_x \right.$$
$$\left. + \frac{1}{Re} \left(\mathcal{D}_N^4 - 2\alpha \mathcal{D}_N^2 + \alpha^2 \mathcal{I} \right) \right) \mathcal{I}_v \hat{v} \quad (4.42)$$

$$\frac{d\hat{\omega}}{dt} = -M_2^{-1} R_2 \tilde{U}_N^{(1)} i\alpha_z \mathcal{I}_v \hat{v} + M_2^{-1} R_2 \left(-\tilde{U}_N^{(0)} i\alpha_x + \frac{1}{Re} \left(\mathcal{D}_N^2 - \alpha \mathcal{I} \right) \right) \mathcal{I}_\omega \hat{\omega}$$
$$(4.43)$$

where $\alpha = \alpha_x^2 + \alpha_z^2$, and

$$\begin{aligned}
M_1 &= R_1 \left(\mathcal{D}_N^2 - \left(\alpha_x^2 + \alpha_z^2 \right) \mathcal{I} \right) \mathcal{I}_v \\
M_2 &= R_2 \mathcal{I}_\omega \\
R_1 &= \begin{bmatrix} 1 & 0 & \cdots & 0 & 0 \\ 0 & 0 & \mathcal{I}_{N-3} & 0 & 0 \\ 0 & 0 & \cdots & 0 & 1 \end{bmatrix} \\
R_2 &= \begin{bmatrix} 0 & \mathcal{I}_{N-1} & 0 \end{bmatrix}.
\end{aligned}$$

The measurement $\hat{z}$ is the following

$$\hat{z}(\alpha_x, \alpha_z, t) = \frac{1}{Re} \begin{bmatrix} -\frac{\partial \hat{u}}{\partial y}(\alpha_x, y = 1, \alpha_z, t) \\ \frac{\partial \hat{u}}{\partial y}(\alpha_x, y = -1, \alpha_z, t) \\ -\frac{\partial \hat{w}}{\partial y}(\alpha_x, y = 1, \alpha_z, t) \\ \frac{\partial \hat{w}}{\partial y}(\alpha_x, y = -1, \alpha_z, t) \end{bmatrix},$$

which is obtained physically by measuring the shear in two directions ($\partial U / \partial y$ and $\partial W / \partial y$) on the entire wall. From continuity (2.55) we have that

$$i\alpha_x \hat{u} + \frac{\partial \hat{v}}{\partial y} + i\alpha_z \hat{w} = 0$$

which, along with $\hat{\omega} = i\alpha_z \hat{u} - i\alpha_x \hat{w}$, yield

$$\begin{aligned}
\hat{u} &= \frac{i}{\alpha} \left(\alpha_x \frac{\partial \hat{v}}{\partial y} - \alpha_z \hat{\omega} \right) \\
\hat{w} &= \frac{i}{\alpha} \left(\alpha_z \frac{\partial \hat{v}}{\partial y} + \alpha_x \hat{\omega} \right).
\end{aligned}$$

So the measurement is

$$\hat{z}(\alpha_x, \alpha_z, t) = \frac{i}{\alpha Re} \begin{bmatrix} -\alpha_x s_{N+1}^T \mathcal{D}_N^2 \mathcal{I}_v \hat{\mathbf{v}} + \alpha_z s_{N+1}^T \mathcal{D}_N^1 \mathcal{I}_\omega \hat{\omega} \\ \alpha_x s_1^T \mathcal{D}_N^{(2)} \mathcal{I}_v \hat{\mathbf{v}} - \alpha_z s_1^T \mathcal{D}_N^1 \mathcal{I}_\omega \hat{\omega} \\ -\alpha_z s_{N+1}^T \mathcal{D}_N^{(2)} \mathcal{I}_v \hat{\mathbf{v}} - \alpha_x s_{N+1}^T \mathcal{D}_N^1 \mathcal{I}_\omega \hat{\omega} \\ \alpha_z s_1^T \mathcal{D}_N^{(2)} \mathcal{I}_v \hat{\mathbf{v}} + \alpha_x s_1^T \mathcal{D}_N^1 \mathcal{I}_\omega \hat{\omega} \end{bmatrix}.$$

Finally, replacing the equations for $\hat{v}(y_0)$ and $\hat{v}(y_N)$ in (4.42) by the control input, we get the system in state space form

$$\dot{\mathbf{x}} = A\mathbf{x} + B\mathbf{u} \tag{4.44}$$

$$z = C\mathbf{x} + D\mathbf{u} \tag{4.45}$$

where

$$\mathbf{x} = \begin{bmatrix} \hat{\mathbf{v}}^T & \hat{\omega}^T \end{bmatrix}^T, \quad \mathbf{u} = \begin{bmatrix} \dot{v}(y_0) & \dot{v}(y_N) \end{bmatrix}^T$$

$$A = \begin{bmatrix} A_{11} & 0 \\ A_{21} & A_{22} \end{bmatrix}, \quad B = \begin{bmatrix} \mathcal{I}_b \\ 0 \end{bmatrix}$$

$$C = \frac{i}{Re\,(\alpha_z^2 + \alpha_x^2)} \begin{bmatrix} -\alpha_x s_{N+1}^T \mathcal{D}_N^2 \mathcal{I}_v & \alpha_z s_{N+1}^T \mathcal{D}_N^1 \mathcal{I}_\omega \\ \alpha_x s_1^T \mathcal{D}_N^{(2)} \mathcal{I}_v & -\alpha_z s_1^T \mathcal{D}_N^1 \mathcal{I}_\omega \\ -\alpha_z s_{N+1}^T \mathcal{D}_N^{(2)} \mathcal{I}_v & -\alpha_x s_{N+1}^T \mathcal{D}_N^1 \mathcal{I}_\omega \\ \alpha_z s_1^T \mathcal{D}_N^{(2)} \mathcal{I}_v & \alpha_x s_1^T \mathcal{D}_N^1 \mathcal{I}_\omega \end{bmatrix}, \quad D = 0_{4 \times 2}$$

$$A_{11} = R_3 M_1^{-1} R_1 \left(\tilde{U}_N^{(0)} \left(-i\alpha_x \mathcal{D}_N^2 + \left(i\alpha_x^3 + i\alpha_x \alpha_z^2\right) I \right) + \tilde{U}_N^{(2)} i\alpha_x \right.$$
$$\left. + \frac{1}{Re} \left(\mathcal{D}_N^4 - 2\left(\alpha_x^2 + \alpha_z^2\right) \mathcal{D}_N^2 + \left(\alpha_x^2 + \alpha_z^2\right)^2 I \right) \right) \mathcal{I}_v$$

$$A_{21} = -M_2^{-1} R_2 \tilde{U}_N^{(1)} i\alpha_z \mathcal{I}_v$$

$$A_{22} = M_2^{-1} R_2 \left(-\tilde{U}_N^{(0)} i\alpha_x + \frac{1}{Re} \left(\mathcal{D}_N^2 - \left(\alpha_x^2 + \alpha_z^2\right) I \right) \right) \mathcal{I}_\omega$$

$$\mathcal{I}_b = \begin{bmatrix} 1 & 0 & \cdots & 0 & 0 \\ 0 & 0 & \cdots & 0 & 1 \end{bmatrix}_{2 \times (N-1)}^T$$

$$R_3 = \begin{bmatrix} 0 & \cdots & 0 \\ \vdots & \mathcal{I}_{N-3} & \vdots \\ 0 & \cdots & 0 \end{bmatrix}.$$

Note that the rigorous derivation of the homogeneous Neumann boundary conditions on v removes the spurious eigenvalues reported in [24], and thus removes the need for redesigning the matrix A to damp out these modes, which is done in that reference.

Control Strategies: A Comparative Study

In [24], two cases are studied in detail: 1) Reynolds number 10000, $\alpha_x = 2$, and $\alpha_z = 0$; and 2) Reynolds number 5000, $\alpha_x = 0$, and $\alpha_z = 2.044$. In case 1, where $\alpha_z = 0$, the equations for ω decouples from the equation for v as well as from the control input, and the problem becomes the same as that studied in the previous sections. Case 2) is a different problem in that it is subcritical, which means that there are no unstable modes. The control problem is nevertheless interesting because perturbations in laminar subcritical flows may lead to transition to turbulence. Therefore, it is of interest to apply control in order to delay, or maybe even prevent, transition to turbulence. The particular case chosen here is the pair (α_x, α_z) that gives the maximum transient energy growth, as shown in [33]. The "worst-case" transient energy growth is defined as

$$\mathcal{E}(t) = \sup_{x_0} \frac{\|x(t)\|_2}{\|x_0\|_2}.$$

The state space model developed in the previous section is subjected to state disturbances, w_1, and measurement noise, w_2, such that it can be written

$$\dot{x} = Ax + B_1 w + B_2 u$$
$$y = C_2 x + D_{21} w$$

where

$$w = \begin{bmatrix} w_1 \\ w_2 \end{bmatrix}, \ B_1 = \begin{bmatrix} I & 0 \end{bmatrix}, \ B_2 = B, \ C_2 = C, \text{ and } D_{21} = \begin{bmatrix} 0 & \alpha I \end{bmatrix}.$$

The performance variable, z, is defined as

$$z = C_1 x + D_{12} u$$

where

$$C_1 = \begin{bmatrix} Q^{1/2} \\ 0 \end{bmatrix}, \ D_{12} = \begin{bmatrix} 0 \\ lI \end{bmatrix}.$$

The system is now on the standard form (3.9)–(3.11), needed for application of the $\mathcal{H}_2$ and $\mathcal{H}_\infty$ control strategies. Q, which shapes the dependence of the performance upon the states, is chosen such that x^*Qx is related to the energy of the flow perturbations, which appears to be the best choice for delaying transition to turbulence [25]. In [24], an extensive parametric study is carried out, quantifying the performance of $\mathcal{H}_2$, $\mathcal{H}_\infty$ and proportional control strategies in terms of the system norms $\|T_{xw}\|_2$, $\|T_{xw}\|_\infty$ and $\|T_{uw}\|_2$, where T_{xw} is the transfer matrix from the disturbance input w to the state x, and T_{uw} is the transfer matrix from the disturbance input w to the control u. Thus, the norms are measures of the state response to Gaussian disturbances, the

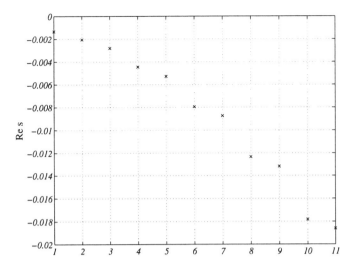

Figure 4.6: The eleven least stable eigenvalues in case 2. All eigenvalues are real [24].

state response to worst-case disturbances, and the control used in response to Gaussian disturbances, respectively. For case 1, the results show that the proportional controllers are not nearly as effective as the $\mathcal{H}_2$ and $\mathcal{H}_\infty$ controllers. The best $\mathcal{H}_\infty$ controller tested, is reported to perform better than all proportional controllers tested with respect to the response of the state to both white noise disturbances and worst case disturbances, and use significantly less control energy than the proportional controllers.

Turning now to case 2, Figure 4.6 shows the eleven least stable poles for this case. Clearly, they are all in the left half plane. Figure 4.7 shows the eigenvectors corresponding to the eleven least stable eigenvalues. The eigenvectors appear pairwise quite similar (except for the first eigenvector), implying that they are highly non-orthogonal. The problem of large transient energy growth is connected with this non-orthogonality, along with the magnitude of the corresponding eigenvalues. This can be illustrated by the following example, involving the second order time invariant system

$$\dot{x} = Ax = \begin{bmatrix} -1 & 0 \\ a & -11 \end{bmatrix} x,$$

with initial condition $x(0) = x_0$. The eigenvalues of A are $\lambda_1 = -1$ and $\lambda_2 = -11$, that is, they are independent of a by the triangular structure of A. The associated normalized eigenvectors of A are $v_1 = [\, 1 \quad a/10 \,]^T / \sqrt{a^2/100 + 1}$ and $v_2 = [\, 0 \quad 1 \,]^T$, respectively. The scalar product of the two eigenvectors is

$$v_1 \cdot v_2 = \frac{a}{\sqrt{a^2 + 100}}$$

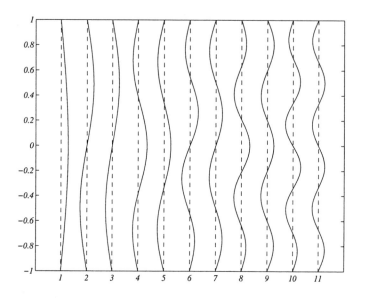

Figure 4.7: Eigenvectors corresponding to the eleven least stable eigenvalues [24].

which is maximized as $a \to \infty$. When $a = 0$ the eigenvectors are orthogonal, and the energy decreases monotonically for all initial conditions, since the solution is simply given by

$$x(t) = \left[\begin{array}{cc} e^{-t} & 0 \\ 0 & e^{-11t} \end{array} \right] x_0$$

in this case. However, if $a = 100$, for instance, the solution is given by

$$x(t) = \left[\begin{array}{cc} e^{-t} & 0 \\ 10 \left(e^{-t} - e^{-11t} \right) & e^{-11t} \end{array} \right] x_0$$

so the (worst case) transient energy growth is

$$\mathcal{E}(t) = \sup_{x_0} \frac{\|x(t)\|_2}{\|x_0\|_2} = \left\| \left[\begin{array}{cc} e^{-t} & 0 \\ 10 \left(e^{-t} - e^{-11t} \right) & e^{-11t} \end{array} \right] \right\|_2 = 10 \left(e^{-t} - e^{-11t} \right),$$

which is plotted in Figure 4.8. When this phenomenon occurs in a channel flow, transition to turbulence is likely to occur, because the nonlinear terms in the Navier-Stokes equation come into play. Thus, it is an important property of any control law for (4.44)–(4.45), that the transient energy growth is suppressed. This is the main point in [24] regarding case 2. The parametric study carried out in [24], indicate that the $\mathcal{H}_2$ and $\mathcal{H}_\infty$ controllers act to make the set of eigenvectors more orthogonal. The maximum transient energy growth of the system is reported to be reduced effectively.

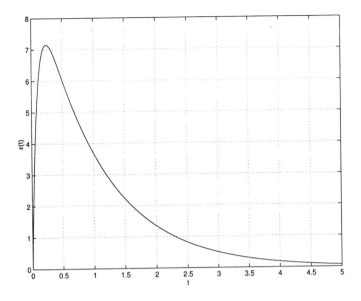

Figure 4.8: Energy growth in second order example system.

4.2   Spatial Invariance Yields Localized Control

In [43], an implementation of the control system for stabilizing the 2D channel flow was suggested. Figure 4.9 shows schematically the setup, involving arrays of sensors and actuators, and a central computing unit. Data from the entire sensor array is fed into the computer, Fourier transformed, and then fed to the control algorithm. The resulting control signal is inverse Fourier transformed and output to the actuator array. The communication needs are tremendous, and so is the computational load. It would be desirable to have localized control, that is, actuation at a certain spatial position should depend on sensing in a neighborhood of that location. Intuitively, flow variables far away from a certain actuation point should be less important than flow variables that are closer. And indeed, this is the case, as shown in [19]. The results are based on the notion of spatial invariance. Loosely put, the system is spatially invariant with respect to the spatial variable x if the system looks the same looking up and down the x-axis, regardless of the point of reference. It is clear that the prototype flows studied here have this property in the streamwise (and spanwise, in the 3D case) direction. In this case, the optimal controllers derived above also are spatially invariant. Moreover, the optimal state feedback $u = K\psi$ at a location x can be written as

$$u(x,t) = \int_{\mathbb{R}} k_s(x - \zeta)\psi(\zeta)\, d\zeta \qquad (4.46)$$

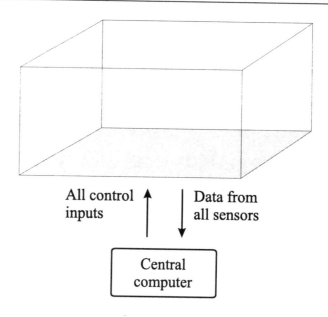

Figure 4.9: Centralized control. Actuators and sensors are distributed over the shaded face. All sensor data is sent to a central computer which calculates and issues control signals to all actuators.

where the convolution kernel k_s decays exponentially, that is

$$|k_s(x)| \leq M e^{-\alpha|x|}$$

for some positive constants M and α. Similarly, an exponentially decaying convolution kernel, k_e, can be found for the estimation problem. Thus, one can approximate the integral (4.46) to any desired degree of accuracy by truncating it at some appropriate $\varepsilon > 0$, to obtain

$$u(x,t) = \int_{\mathbb{R}} k_s(x - \zeta)\psi(\zeta)\,d\zeta \approx \int_{-\varepsilon}^{\varepsilon} k_s(x - \zeta)\psi(\zeta)\,d\zeta.$$

It follows that one can design decentralized controllers that are arbitrarily close to optimal. For the linearized Navier-Stokes equations for 3D channel flow, these kernels have been calculated in [72]. Implementation of the control can be done in terms of a lattice of identical tiles incorporating sensors, actuators and computation logic, as shown schematically in Figure 4.10. Each tile estimates the state above itself, and the information is communicated to its neighbors. Based on gathered information, each tile calculates its control. The choice of ε determines the distance over which sensor information must be communicated.

The implications of the results presented in this section is that the optimal control laws designed for channel flow in the previous sections, which require

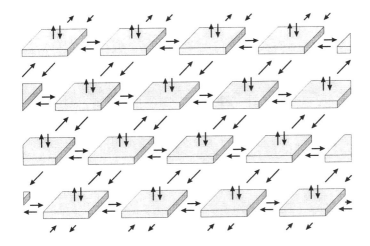

Figure 4.10: Decentralized control. A lattice of identical tiles incorporating sensors and actuators, and computation logic. Sensor information is communicated between neighboring tiles, and each tile computes the control above itself.

huge amounts of wiring and a powerful central computer, can be approximated by localized controllers.

4.3 Lyapunov Stability Approach

In this section, we use Lyapunov stability analysis to show stability of the parabolic equilibrium profiles (2.72) and (2.63), for 2D channel flow and 3D pipe flow, respectively. The results extend easily to 3D channel flow as well. The Navier-Stokes equations are nonlinear, and the only way we can assure global stability of the feedback system is by nonlinear analysis. Note, though, that this does not necessarily mean that the stabilizing feedback control law has to be nonlinear. Indeed, the result of the analysis below, turns out to give astonishingly simple control laws, that are linear and completely decentralized.

4.3.1 2D Channel Flow

Lyapunov Stability Analysis

Boundary control laws for stabilization are sought such that the kinetic energy of the system decays as a function of time. This is a standard Lyapunov-based approach, in which the Lyapunov function is chosen as

$$E(\mathbf{w}) = \|\mathbf{w}\|_{L_2}^2 = \int\limits_{-1}^{1} \int\limits_{0}^{L} \left(u^2 + v^2\right) dx\,dy. \tag{4.47}$$

The Lyapunov analysis is performed in the perturbation variables, since the parabolic equilibrium profile is moved to the origin in these variables. Moving the equilibrium point that is to be stabilized to the origin by means of a coordinate transformation is standard procedure in Lyapunov analysis. The time derivative of $E(\mathbf{w})$ along the trajectories of (2.74)–(2.75) is

$$\dot{E}(\mathbf{w}) = 2 \int\limits_{-1}^{1} \int\limits_{0}^{L} \left(u\frac{\partial u}{\partial t} + v\frac{\partial v}{\partial t}\right) dx\,dy$$

$$= 2 \int\limits_{-1}^{1} \int\limits_{0}^{L} u\left(\frac{1}{Re}\left(\frac{\partial^2 u}{\partial x^2} + \frac{\partial^2 u}{\partial y^2}\right) - u\frac{\partial u}{\partial x} - \tilde{U}\frac{\partial u}{\partial x} - v\frac{\partial u}{\partial y} - v\frac{d\tilde{U}}{dy} - \frac{\partial p}{\partial x}\right) dx\,dy$$

$$+ 2 \int\limits_{-1}^{1} \int\limits_{0}^{L} v\left(\frac{1}{Re}\left(\frac{\partial^2 v}{\partial x^2} + \frac{\partial^2 v}{\partial y^2}\right) - u\frac{\partial v}{\partial x} - \tilde{U}\frac{\partial v}{\partial x} - v\frac{\partial v}{\partial y} - \frac{\partial p}{\partial y}\right) dx\,dy. \tag{4.48}$$

Integration by parts, noticing that $u^2\frac{\partial u}{\partial x} = \frac{1}{2}u\frac{\partial(u^2)}{\partial x}$ and $v^2\frac{\partial v}{\partial y} = \frac{1}{2}v\frac{\partial(v^2)}{\partial y}$, and keeping in mind that all the variables are periodic in x, yields

$$\dot{E}(\mathbf{w}) = \frac{2}{Re} \int\limits_{0}^{L} \left[\frac{\partial u}{\partial y}u + \frac{\partial v}{\partial y}v\right]_{y=-1}^{1} dx$$

$$- \frac{2}{Re} \int\limits_{-1}^{1} \int\limits_{0}^{L} \left(\left(\frac{\partial u}{\partial x}\right)^2 + \left(\frac{\partial u}{\partial y}\right)^2 + \left(\frac{\partial v}{\partial x}\right)^2 + \left(\frac{\partial v}{\partial y}\right)^2\right) dx\,dy$$

$$+ \underbrace{\int\limits_{-1}^{1} \int\limits_{0}^{L} u^2\frac{\partial u}{\partial x}dx\,dy - 2\int\limits_{-1}^{1} \int\limits_{0}^{L} \tilde{U}\frac{\partial u}{\partial x}u\,dx\,dy - \int\limits_{0}^{L} [u^2v]_{y=-1}^{1}\,dx + \int\limits_{-1}^{1} \int\limits_{0}^{L} u^2\frac{\partial v}{\partial y}dx\,dy}_{=0^1}$$

$$- 2\int\limits_{-1}^{1} \int\limits_{0}^{L} vu\frac{d\tilde{U}}{dy}dx\,dy - 2\int\limits_{-1}^{1} \int\limits_{0}^{L} \frac{\partial u}{\partial x}p\,dx\,dy + \int\limits_{-1}^{1} \int\limits_{0}^{L} v^2\frac{\partial u}{\partial x}dx\,dy$$

---

[1]Due to periodic boundary conditions in the streamwise direction: $\int\limits_{0}^{L} \tilde{U}\frac{\partial u}{\partial x}u\,dx = \frac{\tilde{U}}{2}\int\limits_{0}^{L} \frac{\partial(u^2)}{\partial x}dx = \frac{\tilde{U}}{2}[u^2]_0^L = 0.$

$$- 2 \underbrace{\int_{-1}^{1} \int_{0}^{L} \tilde{U} \frac{\partial v}{\partial x} v \, dx \, dy - \int_{0}^{L} [v^3]_{y=-1}^{1} \, dx}_{=0^1} + \int_{-1}^{1} \int_{0}^{L} v^2 \frac{\partial v}{\partial y} \, dx \, dy$$

$$- 2 \int_{0}^{L} [vp]_{y=-1}^{1} \, dx + 2 \int_{-1}^{1} \int_{0}^{L} \frac{\partial v}{\partial y} p \, dx \, dy$$

$$= - \frac{2}{Re} \int_{-1}^{1} \int_{0}^{L} \left(\left(\frac{\partial u}{\partial x} \right)^2 + \left(\frac{\partial u}{\partial y} \right)^2 + \left(\frac{\partial v}{\partial x} \right)^2 + \left(\frac{\partial v}{\partial y} \right)^2 \right) dx \, dy$$

$$- 2 \int_{-1}^{1} \int_{0}^{L} vu \frac{d\tilde{U}}{dy} dx \, dy + \int_{-1}^{1} \int_{0}^{L} u^2 \left(\frac{\partial u}{\partial x} + \frac{\partial v}{\partial y} \right) dx \, dy - 2 \int_{-1}^{1} \int_{0}^{L} p \left(\frac{\partial u}{\partial x} + \frac{\partial v}{\partial y} \right) dx \, dy$$

$$+ \int_{-1}^{1} \int_{0}^{L} v^2 \left(\frac{\partial u}{\partial x} + \frac{\partial v}{\partial y} \right) dx \, dy + \frac{2}{R} \int_{0}^{L} \left[\frac{\partial u}{\partial y} u \right]_{y=-1}^{1} dx$$

$$+ \frac{2}{Re} \int_{0}^{L} \left[\left(\frac{\partial v}{\partial y} - Rp \right) v \right]_{y=-1}^{1} dx - \int_{0}^{L} [(u^2 + v^2) v]_{y=-1}^{1} dx. \quad (4.49)$$

Using continuity (2.73), we get

$$\dot{E}(\mathbf{w}) = - \frac{2}{Re} \int_{-1}^{1} \int_{0}^{L} \left(\left(\frac{\partial u}{\partial x} \right)^2 + \left(\frac{\partial u}{\partial y} \right)^2 + \left(\frac{\partial v}{\partial x} \right)^2 + \left(\frac{\partial v}{\partial y} \right)^2 \right) dx \, dy$$

$$- 2 \int_{-1}^{1} \int_{0}^{L} uv \frac{d\tilde{U}}{dy} dx \, dy + \frac{2}{Re} \int_{0}^{L} \left[\frac{\partial u}{\partial y} u \right]_{y=-1}^{1} dx$$

$$+ \frac{2}{Re} \int_{0}^{L} \left[\left(\frac{\partial v}{\partial y} - Rp \right) v \right]_{y=-1}^{1} dx - \int_{0}^{L} [(u^2 + v^2) v]_{y=-1}^{1} dx. \quad (4.50)$$

Following [16, Lemma 6.2] (also in [15, Lemma 3.2]), we set

$$u(x, y, t) = u(x, -1, t) + \int_{-1}^{y} \frac{\partial u}{\partial y} (x, \gamma, t) d\gamma \quad (4.51)$$

where the integration variable is denoted γ for notational clarity. Squaring (4.51) yields

$$u^2(x, y, t) = \left(u(x, -1, t) + \int_{-1}^{y} \frac{\partial u}{\partial y} (x, \gamma, t) d\gamma \right)^2 \leq$$

$$2u^2(x,-1,t) + 2\left(\int_{-1}^{y} \frac{\partial u}{\partial y}(x,\gamma,t)d\gamma\right)^2.$$

By the Schwartz inequality,

$$\left(\int_{-1}^{y} 1\frac{\partial u}{\partial y}(x,\gamma,t)d\gamma\right)^2 \le (y+1)\left(\int_{-1}^{y}\left(\frac{\partial u}{\partial y}(x,\gamma,t)\right)^2 d\gamma\right) \qquad (4.52)$$

so we have that

$$u^2(x,y,t) \le 2u^2(x,-1,t) + 2(y+1)\int_{-1}^{1}\left(\frac{\partial u}{\partial y}(x,y,t)\right)^2 dy \qquad (4.53)$$

where we have set $y = 1$ in the integral. Therefore, we get

$$\int_{-1}^{1}\int_{0}^{L} u^2 dxdy \le 4\int_{0}^{L} u^2(x,-1,t)dxdy$$

$$+ 2\int_{-1}^{1}\int_{0}^{L}(y+1)\left(\int_{-1}^{1}\left(\frac{\partial u}{\partial y}(x,y,t)\right)^2 dy\right)dxdy$$

$$= 4\int_{0}^{L} u^2(x,-1,t)dxdy + 2\int_{0}^{L}\left(\int_{-1}^{1}(y+1)\,dy\right)\left(\int_{-1}^{1}\left(\frac{\partial u}{\partial y}(x,y,t)\right)^2 dy\right)dx$$

$$= 4\int_{0}^{L} u^2(x,-1,t)dxdy + 4\int_{-1}^{1}\int_{0}^{L}\left(\frac{\partial u}{\partial y}(x,y,t)\right)^2 dxdy. \qquad (4.54)$$

An analogous derivation for $\frac{\partial v}{\partial y}$ now gives

$$-\int_{-1}^{1}\int_{0}^{L}\left(\left(\frac{\partial u}{\partial y}\right)^2 + \left(\frac{\partial v}{\partial y}\right)^2\right)dxdy \le -\frac{E(\mathbf{w})}{4}$$

$$+ \int_{0}^{L}\left(u^2(x,-1,t) + v^2(x,-1,t)\right)dx. \qquad (4.55)$$

Inserting (4.55) into (4.50) we get

$$
\dot{E}(\mathbf{w}) \leq -\frac{1}{2Re} E(\mathbf{w}) + \frac{2}{Re} \int_0^L \left(u^2(x,-1,t) + v^2(x,-1,t) \right) dx
$$

$$
- \frac{2}{Re} \int_{-1}^1 \int_0^L \left(\left(\frac{\partial u}{\partial x} \right)^2 + \left(\frac{\partial v}{\partial x} \right)^2 \right) dx dy
$$

$$
- 2 \int_{-1}^1 \int_0^L uv \frac{d\tilde{U}}{dy} dx dy + \frac{2}{Re} \int_0^L \left[\frac{\partial u}{\partial y} u \right]_{y=-1}^1 dx
$$

$$
+ \frac{2}{Re} \int_0^L \left[\left(\frac{\partial v}{\partial y} - Rp \right) v \right]_{y=-1}^1 dx - \int_0^L \left[(u^2 + v^2) v \right]_{y=-1}^1 dx. \quad (4.56)
$$

Since

$$
-2 \int_{-1}^1 \int_0^L uv \frac{d\tilde{U}}{dy} dx dy \leq 2 \int_{-1}^1 \int_0^L 2\,|u|\,|v|\,dx dy \leq 2 \int_{-1}^1 \int_0^L (u^2 + v^2)\,dx dy = 2E(\mathbf{w})
$$

$$
(4.57)
$$

we finally get

$$
\dot{E}(\mathbf{w}) \leq -\frac{1}{2} \left(\frac{1}{Re} - 4 \right) E(\mathbf{w}) + \frac{2}{Re} \int_0^L \left(u^2(x,-1,t) + v^2(x,-1,t) \right) dx
$$

$$
+ \frac{2}{Re} \int_0^L \left[\frac{\partial u}{\partial y} u \right]_{y=-1}^1 dx + 2 \int_0^L \left[\left(\frac{1}{Re} \frac{\partial v}{\partial y} - p \right) v \right]_{y=-1}^1 dx
$$

$$
- \int_0^L \left[(u^2 + v^2) v \right]_{y=-1}^1 dx. \quad (4.58)
$$

Notice that for $Re < 1/4$, $E(\mathbf{w})$ decays exponentially with time even in the uncontrolled case ($u(x,\pm 1,t) = v(x,\pm 1,t) \equiv 0$). In other words, the fixed point $(\tilde{U}, \tilde{V})$ is globally exponentially stable (in L_2) in this case, and the goal of applying boundary control is to enhance stability. The four last terms in (4.58) are evaluated on the boundary, and are the means by which boundary control laws are designed. Below, two control laws are presented: the first uses wall-tangential actuation; and the second uses wall-normal actuation.

Wall-Tangential Distributed Actuation

The following boundary control was suggested in [16]:

$$u(x,-1,t) \;=\; k_u \frac{\partial u}{\partial y}(x,-1,t), \;\; u(x,1,t) = -k_u \frac{\partial u}{\partial y}(x,1,t) \quad (4.59)$$

$$v(x,-1,t) \;=\; v(x,1,t) = 0. \quad (4.60)$$

Inserting (4.59)–(4.60) into (4.58) gives

$$\dot{E}(\mathbf{w}) \leq -\frac{1}{2}\left(\frac{1}{Re}-4\right)E(\mathbf{w}) - \frac{2}{Re}\left(\frac{1}{k_u}-1\right)\int_0^L u^2(x,-1,t)dx. \quad (4.61)$$

Thus, for $Re < 1/4$ and $k_u \in [0,1]$, $E(\mathbf{w})$ decays exponentially with time.

Wall-Normal Distributed Actuation

Actuation normal to the wall is another strategy of active interest. The inequality (4.58) also suggests a control law structure for wall-normal control [1] (also in [2]). Setting $u(x,-1,t) = u(x,1,t) = 0$, $\frac{\partial v}{\partial y}$ is zero at the wall by continuity (2.73), so we have

$$\dot{E}(\mathbf{w}) \leq -\frac{1}{2}\left(\frac{1}{Re}-4\right)E(\mathbf{w}) + \frac{2}{Re}\int_0^L v^2(x,-1,t)dx$$

$$-2\int_0^L [pv]_{y=-1}^1 \, dx - \int_0^L [v^3]_{y=-1}^1 \, dx. \quad (4.62)$$

Now, by imposing $v(x,-1,t) = v(x,1,t)$, the last term in (4.62) vanishes. Thus, we propose the following control law

$$u(x,-1,t) \;=\; u(x,1,t) = 0 \quad (4.63)$$

$$v(x,-1,t) \;=\; v(x,1,t) = k_v\left(p(x,1,t)-p(x,-1,t)\right). \quad (4.64)$$

Inserting (4.63)–(4.64) into (4.62) gives

$$\dot{E}(\mathbf{w}) \leq -\frac{1}{2}\left(\frac{1}{Re}-4\right)E(\mathbf{w}) - 2\left(\frac{1}{k_v}-\frac{1}{Re}\right)\int_0^L v^2(x,-1,t)dx. \quad (4.65)$$

Thus, for $Re < 1/4$ and $k_v \in [0,Re]$, $E(\mathbf{w})$ decays exponentially with time. Furthermore, note that (4.64) ensures that the net mass flux through the walls be zero.

Implementation

In order to implement the above controllers we have to express them in terms of the actual flow variables, U, V and P. For the wall-tangential case, we get

$$U(x, -1, t) = k_u \left(\frac{\partial U}{\partial y}(x, -1, t) - \frac{d\tilde{U}}{dy}(x, -1, t) \right) \tag{4.66}$$

$$U(x, 1, t) = -k_u \left(\frac{\partial U}{\partial y}(x, 1, t) - \frac{d\tilde{U}}{dy}(x, 1, t) \right) \tag{4.67}$$

$$V(x, -1, t) = V(x, 1, t) = 0, \tag{4.68}$$

and for the wall-normal case we get

$$U(x, -1, t) = U(x, 1, t) = 0 \tag{4.69}$$

$$V(x, -1, t) = V(x, 1, t) = k_v \left(P(x, 1, t) - P(x, -1, t) \right). \tag{4.70}$$

It is interesting to notice that the wall-normal control is independent of the physical parameters of the flow. This is an important property, since the physical parameters of any real flow are subject to inaccuracy. In contrast, $\frac{d\tilde{U}}{dy}(x, \pm 1, t)$ must be known for wall-tangential control.

It is also worth noting that the above control laws are of the Jurdjevic-Quinn [82] type, with respect to the Lyapunov function $E(\mathbf{w})$. This endows these control laws with inverse optimality with respect to a meaningful cost functional, which is in these cases complicated to write.

Numerical Demonstration

The theoretical results in this section are only valid for Reynolds numbers less than $1/4$, for which the parabolic equilibrium profile is globally exponentially stable in the uncontrolled case. Thus, the analysis only tells us that the proposed control laws maintain stability, and not necessarily enhance it. In fact, for wall-normal control, simulations at $Re = 0.1$ show that for $k_v = 0.1$, $E(\mathbf{w})$ converges more slowly to 0 than in the uncontrolled case, whereas for $k_v = -0.1$, stability is enhanced. Although this result was unexpected, it does not contradict the theoretical results.

Being valid for small Reynolds numbers only, the theoretical results are of limited practical value. However, they do suggest controller structures worth testing on flows having higher Reynolds numbers. In [16], results from numerical simulations with wall-tangential control were presented that show stabilization of channel flow at $Re = 15000$. Here, we do a comparison of the performance of the two control laws for flows at $Re = 7500$ and $L = 4\pi$.

The simulations are performed using a hybrid Fourier pseudospectral-finite difference discretization and the fractional step technique based on a hybrid Runge-Kutta/Crank-Nicolson time discretization using the numerical method of [26].

This method is particularly well suited even for the cases with wall-normal actuation because of its implicit treatment of the wall-normal convective terms. The wall-parallel direction is discretized using 128 Fourier-modes, while the wall-normal direction is discretized using energy-conserving central finite differences on a stretched staggered grid with 100 gridpoints. The gridpoints have hyperbolic tangent distribution in the wall-normal direction in order to adequately resolve the high-shear regions near the walls. A fixed flow-rate formulation is used, rather than fixed average pressure gradient, since observations suggest that the approach to equilibrium is faster in this case [77]. The difference between the two formulations is discussed briefly in [118]. The time step is in the range $0.05 - 0.07$ for all simulations. In addition to reporting the time evolution of the energy, $E(\mathbf{w})$, we also consider the instantaneous control effort and drag force as measures of performance. The control effort is defined as

$$C(\mathbf{w}) = \sqrt{\int_0^L \left(|\mathbf{w}(x, -1, t)|^2 + |\mathbf{w}(x, 1, t)|^2 \right) dx} \qquad (4.71)$$

and the drag force as

$$D(\mathbf{w}) = \frac{1}{L} \int_0^L \left(\frac{\partial U}{\partial y}(x, -1, t) - \frac{\partial U}{\partial y}(x, 1, t) \right) dx \qquad (4.72)$$

(Notice that (4.72) is really the mean wall shear, which is related to the drag force by the factor μL). For selected time instants, vorticity maps are also provided. The vorticity, ω, is defined using the actual flow variables (rather than the perturbation variables) as

$$\omega(x, y, t) = \frac{\partial V}{\partial x}(x, y, t) - \frac{\partial U}{\partial y}(x, y, t). \qquad (4.73)$$

A total of six simulations are performed: wall-tangential control with $k_u \in [0.05, 0.1, 0.2]$; and wall-normal control with $k_v \in [-0.125, -0.08, -0.05]$. The parabolic equilibrium profile is unstable for $Re = 7500$, so infinitesimal disturbances will grow, but the flow eventually reaches a statistically steady state, which we call *fully established flow*. For all simulations, the fully established flow, for which $E(\mathbf{w}) \approx 1.3$, is chosen as the initial data. Figure 4.11 shows a vorticity map for the fully established uncontrolled flow. It is similar to vorticity maps presented in [77], and clearly shows the ejection of vorticity from the walls into the core of the channel as described in [77].

Figure 4.12 compares wall-tangential and wall-normal control. It is clear that stabilization is obtained for both controllers, in terms of the energy $E(\mathbf{w})$. Figure 4.12 shows that $E(\mathbf{w})$ decays faster for wall-normal control, and at much less control effort (notice the different scales for $C(\mathbf{w})$ for the two cases in Figure 4.12). The ratio of the peak kinetic energy of the control flow (wall-normal),

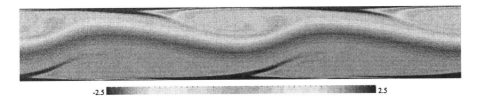

-2.5 ▮▭▭▭▭▭▭▭▮ 2.5

Figure 4.11: Vorticity map for the fully established 2D channel flow (uncontrolled) [1].

versus the perturbation kinetic energy in the uncontrolled case (drained out by the control), $C(\mathbf{w})^2/E(\mathbf{w})$, is less than 0.25%. Also, reduction of drag is more efficient in the wall-normal control case.[1]

Figure 4.13 shows vorticity maps at three different time instances for wall-normal control with $k_v = -0.125$. The removal of vortical structures is evident already at $t = 30$ (top graph), and at $t = 120$ (bottom graph) the flow is nearly uniform. Figure 4.14 shows the pressure field immediately after onset of wall-normal control ($k_v = -0.125$). Regions of low pressure coincide with regions of circulation cells, as the velocity vectors in the intermediate zoom show. In the most detailed zoom, we see that the controller applies suction in this region. This control strategy is discussed further in Section 4.3.5.

In the next section, we will present well posedness results for the case with tangential actuation, along with a numerical simulation that examines that control strategy in more detail.

4.3.2 Regularity of Solutions of the Controlled Channel Flow

In this section, we present results from [16], that provide global stability in H^1 and H^2 norms for the case with wall-tangential actuation introduced in the previous section. The significance of these results is that they rigorously prove that the control input is bounded and go to zero as $t \to \infty$, and that $\mathbf{w}(x, y, t)$ is continuous in all three arguments. This observation has an important practical consequence: the tangential velocity actuation at nearby points on the wall will be in the same direction.

Partial derivatives with respect to x, y, and t will be denoted by corresponding

---

[1] It is interesting to note (see Figure 4.12) that, when the control is applied to the 2D flow, a transient ensues in which the drag dips below the laminar level and then asymptotes towards the laminar state. This transient, however, is dependent on the initial flow state being that of the fully established 2D flow, which has a drag which is significantly higher than laminar. Thus, this transient result does not disprove the conjecture stated in [25], as discussed in Section 4.3.5.

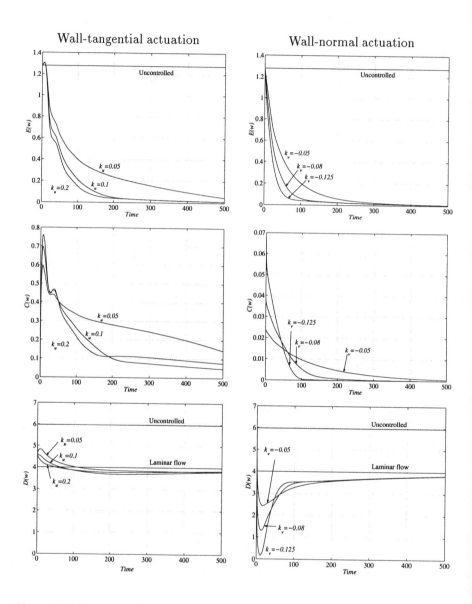

Figure 4.12: Energy $E(\mathbf{w})$ (top row), control effort $C(\mathbf{w})$, and drag $D(\mathbf{w})$ (bottom row), as functions of time for wall-tangential actuation (left column) and wall-normal actuation (right column) [1].

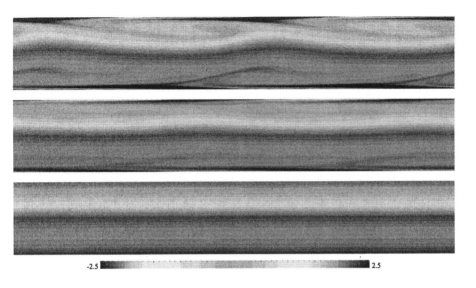

Figure 4.13: Vorticity maps for wall-normal actuation at $t = 30$ (top figure), $t = 60$, and $t = 120$ (bottom figure). The feedback gain is $k_v = -0.125$ [1].

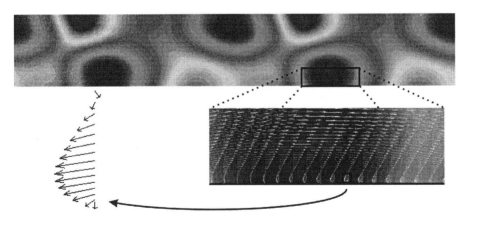

Figure 4.14: Pressure (perturbation only, i.e. p) immediately after onset of wall-normal actuation. Zoom shows velocity vectors in a region with low pressure [1].

subscripts in this section. Recall that for the 2D channel flow, the domain is $\Omega = (0, L) \times (-1, 1)$. For the convenience of the reader, we repeat the equations of motion for 2D channel flow in terms of perturbation variables. They are

$$\begin{cases} u_t - \frac{1}{Re}\Delta u + uu_x + vu_y + \tilde{U}u_x + \tilde{U}'v + p_x = 0, & (x, y) \in \Omega, t > 0, \\ v_t - \frac{1}{Re}\Delta v + uv_x + vv_y + \tilde{U}v_x + p_y = 0, & (x, y) \in \Omega, t > 0, \\ u_x + v_y = 0, & (x, y) \in \Omega, t > 0, \\ u(x, y, 0) = u_0, \, v(x, y, 0) = v_0, & (x, y) \in \Omega, \end{cases}$$

$$(4.74)$$

subject to the periodicity conditions

$$u(0, y, t) = u(L, y, t), \, v(0, y, t) = v(L, y, t), \, -1 < y < 1, t > 0, \quad (4.75)$$

$$v_x(0, y, t) = v_x(L, y, t), \, p(0, y, t) = p(L, y, t), \, -1 < y < 1, t > 0, \quad (4.76)$$

and the boundary control

$$\begin{cases} u(x, -1, t) = k_u u_y(x, -1, t), & 0 < x < L, t > 0, \\ u(x, 1, t) = -k_u u_y(x, 1, t), & 0 < x < L, t > 0, \\ v(x, -1, t) = 0, & 0 < x < L, t > 0, \\ v(x, 1, t) = 0, & 0 < x < L, t > 0, \end{cases}$$

$$(4.77)$$

where k_u is a positive constant.

Mathematical Preliminaries

In what follows, $H^s(\Omega)$ denotes the usual Sobolev space (see [8, 98]) for any $s \in \mathbb{R}$. For $s \geq 0$, $H_0^s(\Omega)$ denotes the completion of $C_0^\infty(\Omega)$ in $H^s(\Omega)$, where $C_0^\infty(\Omega)$ denotes the space of all infinitely differentiable functions on Ω with compact support in Ω. We denote by $\widetilde{H}^s(\Omega)$ the space of the restrictions to Ω of functions which are in $H_{loc}^s(\mathbb{R}^2)$, i.e., $u|_O \in H^s(\Omega)$ for every open bounded set O, and which are periodic in the x–direction:

$$u(x, y) = u(x + L, y). \quad (4.78)$$

The tilde sign will refer to this periodicity in the case of other classical function spaces too.

We shall often be concerned with 2–dimensional vector function spaces and use the following notation to denote them:

$$\widetilde{\mathbf{L}}^2 = \left\{ \widetilde{L}^2(\Omega) \right\}^2, \quad (4.79)$$

$$\widetilde{\mathbf{H}}^1 = \left\{ \widetilde{H}^1(\Omega) \right\}^2, \quad (4.80)$$

$$\widetilde{\mathbf{H}}^2 = \left\{ \widetilde{H}^2(\Omega) \right\}^2, \quad (4.81)$$

$$\tilde{\mathcal{V}} = \left\{ \varphi \in \tilde{C}^{\infty}\left(\Omega\right) : \varphi\left(x, \cdot\right) \in C_0^{\infty}\left(\left(-1, 1\right)\right) \quad \forall x \in [0, L] \right\}, \quad (4.82)$$

$$\tilde{\mathbf{V}} = \left\{ (u, v) \in \tilde{\mathbf{H}}^1 : \begin{array}{c} u_x + v_y = 0 \text{ in } \Omega, \\ v\left(x, -1\right) = v\left(x, 1\right) = 0 \end{array} \right\}, \quad (4.83)$$

$$\tilde{\mathbf{H}} = \text{the closure of } \tilde{\mathbf{V}} \text{ in } \tilde{\mathbf{L}}^2. \quad (4.84)$$

The various norms of these spaces are respectively defined by

$$\|\mathbf{w}\|_{\tilde{\mathbf{L}}^2} = (\mathbf{w}, \mathbf{w})^{1/2}, \quad (4.85)$$

$$\|\mathbf{w}\|_{\tilde{\mathbf{H}}^1} = \left(\|\mathbf{w}\|_{\tilde{\mathbf{L}}^2}^2 + \|\nabla u\|_{\tilde{\mathbf{L}}^2}^2 + \|\nabla v\|_{\tilde{\mathbf{L}}^2}^2\right)^{1/2}, \quad (4.86)$$

$$\|\mathbf{w}\|_{\tilde{\mathbf{H}}^2} = \left(\|\mathbf{w}\|_{\tilde{\mathbf{H}}^1}^2 + \|\nabla u_x\|_{\tilde{\mathbf{L}}^2}^2 + \|\nabla u_y\|_{\tilde{\mathbf{L}}^2}^2 \right. \quad (4.87)$$

$$\left. +\|v_x\|_{\tilde{\mathbf{L}}^2}^2 + \|\nabla v_y\|_{\tilde{\mathbf{L}}^2}^2\right)^{1/2}, \quad (4.88)$$

$$\|\mathbf{w}\|_{\tilde{\mathbf{V}}} = ((\mathbf{w}, \mathbf{w}))^{1/2}, \quad (4.89)$$

where $(\cdot, \cdot)$ denotes the inner product of $\tilde{\mathbf{L}}^2$ and $((\cdot, \cdot))$ denotes the inner product of $\tilde{\mathbf{V}}$ defined by

$$((\mathbf{w}, \mathbf{\Phi})) = \int_{-1}^{1} \int_{0}^{L} Tr\left\{\nabla \mathbf{w}^T \nabla \mathbf{\Phi}\right\} dx dy \quad (4.90)$$

$$+ \frac{1}{k_u} \int_{0}^{L} \left(u\left(x, -1\right)\xi\left(x, -1\right) + u\left(x, 1\right)\xi\left(x, 1\right)\right) dx, \quad (4.91)$$

for all $\mathbf{w} = (u, v)$, $\mathbf{\Phi} = (\xi, \eta) \in \tilde{\mathbf{V}}$.

Let X be a Banach space. We denote by $C^l([0, T]; X)$ the space of l times continuously differentiable functions defined on $[0, T]$ with values in X, and write $C([0, T]; X)$ for $C^0([0, T]; X)$.

Definition 4.1 *A function* $\mathbf{w} = (u, v) \in L^2([0, T]; \tilde{\mathbf{V}})$ *is a* weak solution *of system (4.74)–(4.77) if*

$$\frac{d}{dt}(\mathbf{w}, \mathbf{\Phi}) + \frac{1}{Re}((\mathbf{w}, \mathbf{\Phi})) + ((\mathbf{w} \cdot \nabla)\mathbf{w}, \mathbf{\Phi}) + \left(\tilde{U}\mathbf{w}_x, \mathbf{\Phi}\right) + \left(\tilde{U}'v, \xi\right) = 0 \quad (4.92)$$

is satisfied for all $\mathbf{\Phi} = (\xi, \eta) \in \tilde{\mathbf{V}}$ *and* $\mathbf{w}(x, y, 0) = \mathbf{w}_0(x, y)$ *for all* $(x, y) \in \Omega$.

The Results

Theorem 4.1 *Suppose that*

$$Re < \frac{1}{4} \quad and \quad 0 < k_u < 1, \quad (4.93)$$

and denote

$$\sigma = \frac{1}{4Re} - 1 > 0. \tag{4.94}$$

Then there exists a positive constant $c > 0$ independent of $\mathbf{w}_0$ such that the following statements are true for all $t \geq 0$ for the system (4.74) with periodic conditions (4.75)–(4.76) and boundary control (4.77).

1. *For arbitrary initial data $\mathbf{w}_0(x) \in \widetilde{\mathbf{H}}$, there exists a unique weak solution $\mathbf{w} \in L^2\left([0,\infty); \widetilde{\mathbf{V}}\right) \cap C\left([0,\infty); \widetilde{\mathbf{L}}^2\right)$ that satisfies the following global-exponential stability estimate:*

$$\|\mathbf{w}(t)\| \leq \|\mathbf{w}_0\| e^{-\sigma t}. \tag{4.95}$$

2. *For arbitrary initial data $\mathbf{w}_0(x) \in \widetilde{\mathbf{V}}$, there exists a unique weak solution $\mathbf{w} \in L^2\left([0,\infty); \widetilde{\mathbf{H}}^2 \cap \widetilde{\mathbf{V}}\right) \cap L^\infty\left([0,\infty); \widetilde{\mathbf{V}}\right)$ that satisfies the following global-asymptotic and semiglobal-exponential stability estimate:*

$$\|\mathbf{w}(t)\|_{\widetilde{\mathbf{H}}^1} \leq c\|\mathbf{w}_0\|_{\widetilde{\mathbf{H}}^1} \exp(c\|\mathbf{w}_0\|_{\widetilde{\mathbf{H}}^1}^4)e^{-\sigma t/2}. \tag{4.96}$$

3. *For arbitrary initial data $\mathbf{w}_0(x) \in \widetilde{\mathbf{H}}^2 \cap \widetilde{\mathbf{V}}$ compatible with the control (4.77), there exists a unique weak solution $\mathbf{w} \in C^1\left([0,\infty); \widetilde{\mathbf{L}}^2\right) \cap C\left([0,\infty); \widetilde{\mathbf{H}}^2 \cap \widetilde{\mathbf{V}}\right)$ that satisfies the following global-asymptotic and semiglobal-exponential stability estimate:*

$$\|\mathbf{w}(t)\|_{\widetilde{\mathbf{H}}^2} \leq c\|\mathbf{w}_0\|_{\widetilde{\mathbf{H}}^2} \exp(c\|\mathbf{w}_0\|_{\widetilde{\mathbf{H}}^2}^4)e^{-\sigma t/2}. \tag{4.97}$$

The bound of the form (4.97) also applies to $\|\mathbf{w}_t(t)\|$, $\|\nabla p(t)\|$ and $\max_{(x,y)\in\overline{\Omega}} |\mathbf{w}(x,y,t)|$.

In all of the above cases solutions depend continuously on the initial data in the L^2-norm and the existence, uniqueness and regularity statements hold for any $Re > 0$ and $k_u > 0$ over finite time intervals.

Remark 4.1 Weak solutions satisfying the regularity stated in parts 2 and 3 of Theorem 4.1 are called strong solutions in the literature. Part 3 of Theorem 4.1 means, in particular, that

1. The control inputs $u(x, -1, t)$ and $u(x, 1, t)$ are bounded and go to zero as $t \to \infty$.

2. The regularity statement implies that $\mathbf{w}(x, y, t)$ is continuous in all three arguments. This observation has an important practical consequence: the tangential velocity actuation at nearby points on the wall will be in the same direction.

Remark 4.2 If the Reynolds number $Re \geq 1/4$, the problem of boundary control remains open. The methods presented in this section can not be applied to this case and a radically different method needs to be developed.

Technical Lemmas

In this section, we establish technical lemmas which are the key to proving our main results.

Since $\widetilde{\mathbf{H}}$ is a closed subspace of $\widetilde{\mathbf{L}}^2$, we have the orthogonal decomposition

$$\widetilde{\mathbf{L}}^2 = \widetilde{\mathbf{H}} \oplus \widetilde{\mathbf{H}}^\perp, \tag{4.98}$$

where $\widetilde{\mathbf{H}}^\perp$ denotes the orthogonal complement of $\widetilde{\mathbf{H}}$. Let $\mathcal{P}$ denote the projection from $\widetilde{\mathbf{L}}^2$ onto $\widetilde{\mathbf{H}}$. We define the linear operator A on $\widetilde{\mathbf{H}}$ as

$$A\mathbf{w} = -\mathcal{P}\Delta\mathbf{w}, \tag{4.99}$$

with the domain $D(A)$

$$D(A) = \left\{ \mathbf{w} = (u,v) \in \widetilde{\mathbf{H}} \cap \widetilde{\mathbf{V}} : \begin{array}{c} u(x,-1) = k_u u_y(x,-1), \\ u(x,1) = -k_u u_y(x,1) \end{array} \right\}. \tag{4.100}$$

We first give some basic properties of the subspaces $\widetilde{\mathbf{H}}$, $\widetilde{\mathbf{H}}^\perp$ and the operator A. These properties are similar to the classical results in the cases with homogeneous Dirichlet boundary condition (see, e.g., [127, Chap.I, Sect.1], [39, Chap.4]) and periodic boundary condition (see, e.g., [128, Chap.2]). Thus, their proofs are also similar, however, for completeness, we give brief proofs. The following lemma shows that (4.98) is in fact the so called Helmholtz decomposition of $\widetilde{\mathbf{L}}^2$.

Lemma 4.1 *The subspaces $\widetilde{\mathbf{H}}$ and $\widetilde{\mathbf{H}}^\perp$ can be characterized as follows:*

$$\widetilde{\mathbf{H}}^\perp = \left\{ \mathbf{w} \in \widetilde{\mathbf{L}}^2 : \mathbf{w} = \nabla p, \, p \in \widetilde{H}^1(\Omega) \right\}, \tag{4.101}$$

$$\widetilde{\mathbf{H}} = \left\{ \mathbf{w} = (u,v) \in \widetilde{\mathbf{L}}^2 : \mathrm{div}\mathbf{w} = 0, \, v(x,-1) = v(x,1) = 0 \right\} \tag{4.102}$$

Proof. The proof of (4.102) is the same as the proof of Theorem 1.4 in [127, p.15]. We include the proof of (4.101) which is based on the proof of Theorem 1 in [92, p.27].

Let $\mathbf{w} = (u,v)$ belong to the space on the right hand side of (4.101). Then for all $\mathbf{z} = (\psi, \xi) \in \widetilde{\mathbf{V}}$ we have, using integration by parts,

$$\int_{-1}^{1} \int_{0}^{L} (u\psi + v\xi) \, dxdy = \int_{-1}^{1} \int_{0}^{L} (p_x\psi + p_y\xi) \, dxdy = 0. \tag{4.103}$$

Since $\widetilde{\mathbf{V}}$ is dense in $\widetilde{\mathbf{H}}$, we deduce that $\mathbf{w} \in \widetilde{\mathbf{H}}^{\perp}$.

Conversely, if $\mathbf{w} = (u, v) \in \widetilde{\mathbf{H}}^{\perp}$, then

$$\int_{-1}^{1} \int_{0}^{L} (u\psi + v\xi) \, dxdy = 0, \quad \forall \mathbf{z} = (\psi, \xi) \in \widetilde{\mathbf{V}}. \tag{4.104}$$

Let $\omega_{\rho}(x, y)$ denote a mollifier. For $\varphi \in \widetilde{\mathcal{V}}$, we denote by φ_{ρ} its average:

$$\varphi_{\rho}(x, y) = \int_{\mathbb{R}^{2}} \omega_{\rho}(x - s, y - \tau) \varphi(x, \tau) \, dsd\tau. \tag{4.105}$$

If ρ is small enough, then φ_{ρ} is well–defined on $\Omega_{\rho} = [0, L] \times [-1 + \rho, 1 - \rho]$, it is periodic in the x–direction and vanishes with its derivatives on the horizontal lines $y = -1 + \rho$ and $y = 1 - \rho$. Hence

$$\mathbf{z} = (\varphi_{\rho y}, -\varphi_{\rho x}) \in \widetilde{\mathbf{V}}. \tag{4.106}$$

Thus, we have

$$0 = \int_{-1}^{1} \int_{0}^{L} (u\varphi_{\rho y} - v\varphi_{\rho x}) \, dxdy = \int_{-1}^{1} \int_{0}^{L} (u_{\rho}\varphi_{y} - v_{\rho}\varphi_{x}) \, dxdy$$
$$= \int_{-1}^{1} \int_{0}^{L} (v_{\rho x} - u_{\rho y}) \, dxdy, \tag{4.107}$$

where the functions u_{ρ} and v_{ρ} are defined on Ω_{ρ} and are the averages of u and v respectively. Since $\varphi \in \widetilde{\mathcal{V}}$ is arbitrary and $\widetilde{\mathcal{V}}$ is dense in $\widetilde{L}^{2}(\Omega_{\rho})$, we have

$$v_{\rho x} = u_{\rho y} \quad \text{on } \Omega_{\rho}. \tag{4.108}$$

Take any $y_{0} \in [-1 + \rho, 1 - \rho]$ and define

$$p_{\rho}(x, y) = \int_{(0, y_{0})}^{(x, y)} u_{\rho} \, dx + v_{\rho} \, dy. \tag{4.109}$$

Then we have

$$\mathbf{w}_{\rho} = (u_{\rho}, v_{\rho}) = \nabla p_{\rho} \quad \text{on } \Omega_{\rho}. \tag{4.110}$$

It is well known that for any fixed interior subdomain Ω' of Ω, $\mathbf{w}_{\rho}$ converges to $\mathbf{w}$ in $\widetilde{\mathbf{L}}^{2}(\Omega')$ and then p_{ρ} converges to a function p in $H^{1}(\Omega')$ and

$$\mathbf{w} = \nabla p \quad \text{on } \Omega'. \tag{4.111}$$

Since Ω' is arbitrary, we have

$$\mathbf{w} = \nabla p \quad \text{on } \Omega. \tag{4.112}$$

Finally, we show that p is periodic in the x–direction. Let $\mathbf{z}(x,y) = (\psi(y), 0)$, where $\psi \in C_0^\infty([-1,1])$. Clearly $\mathbf{z} \in \widetilde{\mathbf{V}}$, and

$$0 = \int_{-1}^1 \int_0^L \mathbf{w}_\rho \cdot \mathbf{z} \, dx\, dy = \int_{-1}^1 \int_0^L u_\rho(x,y)\, \psi(y)\, dx\, dy \, \forall y \in [-1,1]. \quad (4.113)$$

Since ψ is from a dense subset of L^2, we obtain

$$\int_0^L u_\rho(x,y)\, dx = 0 \qquad \text{for} \qquad (4.114)$$

With this and with definition (4.109) we obtain that p_ρ, and hence p is periodic in the x–direction.

∎

Lemma 4.2 *The norm* $\|\mathbf{w}\|_{\widetilde{\mathbf{V}}}$ *on* $\widetilde{\mathbf{V}}$ *is equivalent to the norm* $\|\mathbf{w}\|_{\widetilde{\mathbf{H}}^1}$ *induced by* $\widetilde{\mathbf{H}}^1$.

Proof. Using the identity

$$u(x,y) = u(x,-1) + \int_{-1}^y u_y(x,y)\, dy, \quad (4.115)$$

we have

$$\int_{-1}^1 \int_0^L u^2 \, dx\, dy \leq 4 \int_0^L u^2(x,-1)\, dx + 4 \int_{-1}^1 \int_0^L u_y^2 \, dx\, dy. \quad (4.116)$$

Similarly, we have

$$\int_{-1}^1 \int_0^L v^2 \, dx\, dy \leq 2 \int_{-1}^1 \int_0^L v_y^2 \, dx\, dy. \quad (4.117)$$

It therefore follows that

$$\int_{-1}^1 \int_0^L (u^2 + v^2) \, dx\, dy \leq 4 \int_0^L u^2(x,0)\, dx + 4 \int_{-1}^1 \int_0^L (u_y^2 + v_y^2) \, dx\, dy, \quad (4.118)$$

which shows that

$$\|\mathbf{w}\|_{\widetilde{\mathbf{H}}^1} \leq c\|\mathbf{w}\|_{\widetilde{\mathbf{V}}}. \quad (4.119)$$

On the other hand, using (4.115) again, we deduce that

$$\int_0^L u^2(x,-1)\, dx \leq c \int_{-1}^1 \int_0^L (u^2 + u_y^2) \, dx\, dy. \quad (4.120)$$

Similarly, we have

$$\int_0^L u^2(x,1)\,dx \le c \int_{-1}^1 \int_0^L \left(u^2 + u_y^2\right)\,dxdy\,. \qquad (4.121)$$

It therefore follows that

$$\|\mathbf{w}\|_{\widetilde{\mathbf{V}}} \le c\|\mathbf{w}\|_{\widetilde{\mathbf{H}}^1}\,. \qquad (4.122)$$

■

Lemma 4.3 *The norm* $\|A\mathbf{w}\|$ *on* $D(A)$ *is equivalent to the norm* $\|\mathbf{w}\|_{\widetilde{\mathbf{H}}^2}$ *induced by* $\widetilde{\mathbf{H}}^2$.

Proof. By the definition of the operator A, we have

$$(A\mathbf{w}\,\boldsymbol{\Phi}) = ((\mathbf{w}\,\boldsymbol{\Phi}))\,, \qquad \forall \mathbf{w} = (u,v) \in D(A)\,, \quad \boldsymbol{\Phi} = (\xi,\eta) \in \widetilde{\mathbf{V}}\,. \qquad (4.123)$$

As in the proof of regularity of solutions of the Stokes equations with homogeneous Dirichlet boundary conditions (see, e.g., [39, Chap.3]), we can readily prove that

$$D(A) = \left\{ \mathbf{w} \in \widetilde{\mathbf{H}} \,:\, A\mathbf{w} \in \widetilde{\mathbf{H}} \right\}\,. \qquad (4.124)$$

Hence, by Proposition 9 of [46, p.370], $D(A)$ is a Banach space when provided with the graph norm

$$\|\mathbf{w}\|_{D(A)} = \left(\|\mathbf{w}\|^2 + \|A\mathbf{w}\|^2\right)^{1/2}\,.$$

In addition, $D(A)$ with the norm $\|\cdot\|_{\widetilde{\mathbf{H}}^2}$ is also a Banach space, and the norm $\|\cdot\|_{\widetilde{\mathbf{H}}^2}$ is stronger than $\|\cdot\|_{D(A)}$. By the Banach open mapping theorem (see, e.g., [119, p.49]), these two norms $\|\mathbf{w}\|_{\widetilde{\mathbf{H}}^2}$ and $\|\mathbf{w}\|_{D(A)}$ on $D(A)$ are equivalent. On the other hand, by (4.118), we have

$$\|\mathbf{w}\| \le c\,\|A\mathbf{w}\|\,. \qquad (4.125)$$

Hence, the norm $\|A\mathbf{w}\|$ is equivalent to the norm $\|\mathbf{w}\|_{D(A)}$, and then equivalent to the norm induced by $\widetilde{\mathbf{H}}^2$.

■

The following inequality is a special 2–dimensional extension of a classical inequality (see, e.g. [93])

$$\|\varphi\|_{L^q} \le \beta \|\nabla\varphi\|_{L^m}^\alpha \|\varphi\|_{L^r}^{1-\alpha}\,, \qquad (4.126)$$

which holds for any $\varphi \in \overset{\circ}{W}_m^1 (\Omega)$, $m \geq 2$, $r \geq 1$, where $\Omega \subset \mathbb{R}^2$, $r \leq q < \infty$,

$$\alpha = \left(\frac{1}{r} - \frac{1}{q}\right) \left(\frac{1}{2} - \frac{1}{m} + \frac{1}{r}\right)^{-1}$$

and

$$\beta = \max \left\{\frac{q}{2}; 1 + (m-1)mr\right\}.$$

Here $\overset{\circ}{W}_m^1$ denotes the subspace of $L^m (\Omega)$ functions whose gradient is also in $L^m (\Omega)$ and in which the set $C_0^\infty (\Omega)$ is dense.

Lemma 4.4 *For any rectangular region* $\Omega = [0, l] \times [0, k] \subset \mathbb{R}^2$, *where* $k, l > 0$, *and for any* $\varphi \in H^1 (\Omega)$ *and* $2 \leq q < \infty$ *the following inequality holds:*

$$\|\varphi\|_{L^q} \leq \gamma_1 \|\varphi\| + \gamma_2 \|\nabla\varphi\|^\alpha \|\varphi\|^{1-\alpha}, \tag{4.127}$$

where $\alpha = 1 - 2/q$ *and* γ_1, γ_2 *are positive constants depending only on the size of* Ω *and on* q.

Proof. Consider an arbitrary $\varphi \in H^1 (\Omega)$ and its extension

$$\widetilde{\varphi}(x, y) = \begin{cases} \varphi(x, y) & \text{if } (x, y) \in [0, l] \times [0, k], \\ \left(1 + \frac{x}{l}\right) \varphi(-x, y) & \text{if } (x, y) \in [-l, 0] \times [0, k], \\ \left(2 - \frac{x}{l}\right) \varphi(2l - x, y) & \text{if } (x, y) \in [l, 2l] \times [0, k], \\ \left(1 + \frac{y}{k}\right) \varphi(x, -y) & \text{if } (x, y) \in [0, l] \times [-k, 0], \\ \left(2 - \frac{y}{k}\right) \varphi(x, 2k - y) & \text{if } (x, y) \in [0, l] \times [k, 2k], \\ \left(1 + \frac{x}{l}\right)\left(1 + \frac{y}{k}\right) \varphi(-x, -y) & \text{if } (x, y) \in [-l, 0] \times [-k, 0], \\ \left(1 + \frac{x}{l}\right)\left(2 - \frac{y}{k}\right) \varphi(-x, 2k - y) & \text{if } (x, y) \in [-l, 0] \times [k, 2k], \\ \left(2 - \frac{x}{l}\right)\left(1 + \frac{y}{k}\right) \varphi(2l - x, -y) & \text{if } (x, y) \in [l, 2l] \times [-k, 0], \\ \left(2 - \frac{x}{l}\right)\left(2 - \frac{y}{k}\right) \varphi(2l - x, 2k - y) & \text{if } (x, y) \in [0, l] \times [k, 2k], \end{cases} \tag{4.128}$$

Inequality (4.126) applies to $\widetilde{\varphi}$ with $\alpha = 1 - 2/q$ and $r = 2 \leq q < \infty$, since $\widetilde{\varphi} \in H^1\left(\widetilde{\Omega}\right)$ and $\widetilde{\varphi}(x, y) = 0$ for $(x, y) \in \partial\widetilde{\Omega}$, where $\widetilde{\Omega} = [-l, 2l] \times [-k, 2k]$. We have

$$\|\widetilde{\varphi}\|_{L^q(\widetilde{\Omega})} \leq \beta \|\nabla\widetilde{\varphi}\|_{L^2(\widetilde{\Omega})}^\alpha \|\widetilde{\varphi}\|_{L^2(\widetilde{\Omega})}^{1-\alpha}. \tag{4.129}$$

We have the following relationships between the norms of $\widetilde{\varphi}$ and φ.

$$\|\varphi\|_{L^q(\Omega)} \le \|\widetilde{\varphi}\|_{L^q(\widetilde{\Omega})} , \tag{4.130}$$

$$\|\widetilde{\varphi}\|^2_{L^2(\widetilde{\Omega})} \le 9\,\|\varphi\|^2 \tag{4.131}$$

and

$$\|\nabla\widetilde{\varphi}\|^2_{L^2(\widetilde{\Omega})} \le 17\,\|\nabla\varphi\|^2 + 6\left(\frac{2}{l^2} + \frac{2}{k^2}\right)\|\varphi\|^2 . \tag{4.132}$$

Inequality (4.130) and (4.131) are trivial consequences of definition (4.128). In order to see the validity of (4.132) one has to estimate the different pieces of $\nabla\widetilde{\varphi}$. One of these estimates, for example is the following:

$$\int_l^{2l}\int_k^{2k} \left|\nabla\left(\left(2 - \frac{x}{l}\right)\left(2 - \frac{y}{k}\right)\varphi\left(2l - x, 2k - y\right)\right)\right|^2 dx\,dy =$$

$$\int_l^{2l}\int_k^{2k} \left(\frac{1}{l}\left(2 - \frac{y}{k}\right)\varphi\left(2l - x, 2k - y\right)\right.$$

$$\left. + \left(2 - \frac{x}{l}\right)\left(2 - \frac{y}{k}\right)\varphi_x\left(2l - x, 2k - y\right)\right)^2 dx\,dy$$

$$+ \int_l^{2l}\int_k^{2k} \left(\frac{1}{k}\left(2 - \frac{x}{l}\right)\varphi\left(2l - x, 2k - y\right)\right.$$

$$\left. + \left(2 - \frac{x}{l}\right)\left(2 - \frac{y}{k}\right)\varphi_y\left(2l - x, 2k - y\right)\right)^2 dx\,dy$$

$$\le \frac{2}{l^2}\,\|\varphi\|^2 + 2\,\|\varphi_x\|^2 + \frac{2}{k^2}\,\|\varphi\|^2 + 2\,\|\varphi_y\|^2$$

$$= \left(\frac{2}{l^2} + \frac{2}{k^2}\right)\|\varphi\|^2 + 2\,\|\nabla\varphi\|^2 . \tag{4.133}$$

Combining inequalities (4.129)–(4.132) we obtain

$$\|\varphi\|_{L^q(\Omega)} \le \beta\left(17\,\|\nabla\varphi\|^2 + 6\left(\frac{2}{l^2} + \frac{2}{k^2}\right)\|\varphi\|^2\right)^{\frac{\alpha}{2}} 9^{\frac{1-\alpha}{2}}\,\|\varphi\|^{1-\alpha}$$

$$\le \beta\left(17^{\frac{\alpha}{2}}\,\|\nabla\varphi\|^\alpha + \left(\frac{12}{l^2} + \frac{12}{k^2}\right)^{\frac{\alpha}{2}}\|\varphi\|^\alpha\right) 9^{\frac{1-\alpha}{2}}\,\|\varphi\|^{1-\alpha}$$

$$= \gamma_1\,\|\varphi\| + \gamma_2\,\|\nabla\varphi\|^\alpha\,\|\varphi\|^{1-\alpha} . \tag{4.134}$$

∎

Proof of Theorem

We first establish our a priori stability estimates and then deal with questions of existence, uniqueness and regularity.

As in the previous section, we define the perturbation energy $E(\mathbf{w})$ as

$$E(\mathbf{w}) = \|\mathbf{w}\|^2 = \int_{-1}^{1} \int_{0}^{L} \left(u^2 + v^2\right) \, dx\,dy. \tag{4.135}$$

In addition, we define the high order energy $J(\mathbf{w})$ of (4.74)–(4.77) as

$$J(\mathbf{w}) = \|\mathbf{w}\|_{\tilde{V}}^2 = \int_{-1}^{1} \int_{0}^{L} \left(u_x^2 + u_y^2 + v_x^2 + v_y^2\right) \, dx\,dy$$

$$+ \frac{1}{k_u} \int_{0}^{L} \left(u^2(x,-1) + u^2(x,1)\right) \, dx. \tag{4.136}$$

Part 1. The time derivative of $E(\mathbf{w})$ along solutions of (4.74)–(4.77) was calculated in the previous section. In the case when $v|_{wall} = 0$, the result was

$$\dot{E}(\mathbf{w}) = -\frac{2}{Re} \int_{-1}^{1} \int_{0}^{L} \left(u_x^2 + u_y^2 + v_x^2 + v_y^2\right) \, dx\,dy - 2 \int_{-1}^{1} \int_{0}^{L} \tilde{U}' uv \, dx\,dy$$

$$+ \frac{2}{Re} \int_{0}^{L} u_y u \big|_{y=-1}^{1} \, dx. \tag{4.137}$$

It therefore follows from (4.118) that

$$\dot{E}(\mathbf{w}) \leq -\frac{1}{2Re} E(\mathbf{w}) + \frac{2}{Re} \int_{0}^{L} u^2(x,-1,t) \, dx + 2E(\mathbf{w})$$

$$- \frac{2}{k_u Re} \int_{0}^{L} \left(u^2(x,1,t) + u^2(x,-1,t)\right) \, dx$$

$$= -\frac{1}{2Re} E(\mathbf{w}) + 2E(\mathbf{w})$$

$$- \int_{0}^{L} \left(\frac{2}{Re}\left(\frac{1}{k_u} - 1\right) u^2(x,-1,t) + \frac{2}{k_u Re} u^2(x,1,t)\right) \, dx$$

$$\leq -2\left(\frac{1}{4Re} - 1\right) E(\mathbf{w}). \tag{4.138}$$

This implies (4.95).

Part 2. By (4.118) and (4.137), we have

$$\dot{E}(\mathbf{w}) \leq -\frac{2}{Re} \int_{-1}^{1} \int_{0}^{L} \left(u_x^2 + u_y^2 + v_x^2 + v_y^2\right) \, dx\,dy + 2E(\mathbf{w})$$

$$- \frac{2}{k_u Re} \int_{0}^{L} \left(u^2(x,1,t) + u^2(x,-1,t)\right) \, dx$$

$$\leq -2\left(\frac{1}{Re} - 4\right) \int_{-1}^{1} \int_{0}^{L} \left(u_x^2 + u_y^2 + v_x^2 + v_y^2\right) \, dx\,dy$$

$$-\int_0^L \left(2 \left(\frac{1}{k_u Re} - 4 \right) u^2 (x, -1, t) + \frac{2}{k_u Re} u^2 (x, 1, t) \right) dx$$
$$\leq -cJ(\mathbf{w}),$$
(4.139)

where, by (4.93)

$$c = 2 \left(\frac{1}{Re} - 4 \right) > 0.$$
(4.140)

Multiplying (4.139) by $e^{\sigma t}$, we obtain

$$\frac{d}{dt} \left(e^{\sigma t} E(\mathbf{w}) \right) + c e^{\sigma t} J(\mathbf{w}) \leq \sigma e^{\sigma t} E(\mathbf{w}) \leq \sigma E(\mathbf{w}_0) e^{-\sigma t}.$$
(4.141)

Integrating from 0 to t gives

$$e^{\sigma t} E(\mathbf{w}(t)) + c \int_0^t e^{\sigma s} J(\mathbf{w}(s)) \, ds \leq E(\mathbf{w}_0) \left(2 - e^{-\sigma t} \right),$$
(4.142)

which implies

$$c \int_0^t e^{\sigma s} J(\mathbf{w}(s)) \, ds \leq 2E(\mathbf{w}_0), \qquad \forall t \geq 0.$$
(4.143)

In order to obtain further estimates on J, we multiply the first equation of (4.74) by Au and the second equation of (4.74) by Av and integrate over Ω by parts. This gives

$$\int_{-1}^1 \int_0^L (u_t Au + v_t Av) \, dx dy = \frac{1}{Re} \int_{-1}^1 \int_0^L (\Delta u Au + \Delta v Av) \, dx dy$$
$$- \int_{-1}^1 \int_0^L \left(u u_x + v u_y + \tilde{U} u_x + \tilde{U}' v + p_x \right) Au \, dx dy$$
$$- \int_{-1}^1 \int_0^L \left(u v_x + v v_y + \tilde{U} v_x + p_y \right) Av \, dx dy.$$
(4.144)

Since there exists $\mathbf{z} \in \tilde{\mathbf{H}}^\perp$ such that

$$\Delta \mathbf{w} = \mathcal{P} \Delta \mathbf{w} + \mathbf{z},$$
(4.145)

we have (noting that $\int_{-1}^1 \int_0^L \mathbf{w}_t \cdot \mathbf{z} \, dx dy = 0$)

$$\int_{-1}^1 \int_0^L (u_t Au + v_t Av) \, dx dy = \int_{-1}^1 \int_0^L (u_t \Delta u + v_t \Delta v - \mathbf{w}_t \cdot \mathbf{z}) \, dx dy$$
$$= \int_{-1}^1 (u_t u_x + v_t v_x)|_{x=0}^L \, dy + \int_0^L (u_t u_y + v_t v_y)|_{y=-1}^1 \, dx$$
$$- \int_{-1}^1 \int_0^L (u_{xt} u_x + u_{yt} u_y + v_{xt} v_x + v_{yt} v_y) \, dx dy$$
$$= -\frac{1}{2} \dot{J}(\mathbf{w}),$$
(4.146)

and (noting that $\int_{-1}^{1} \int_{0}^{L} A\mathbf{w} \cdot \mathbf{z} \, dx dy = 0$)

$$\int_{-1}^{1} \int_{0}^{L} (\Delta u Au + \Delta v Av) \, dx dy = \int_{-1}^{1} \int_{0}^{L} \|A\mathbf{w}\|^2 \, dx dy . \tag{4.147}$$

Moreover, since $A\mathbf{w} \in \tilde{\mathbf{H}}$ and $\nabla p \in \tilde{\mathbf{H}}^\perp$, we have

$$\int_{-1}^{1} \int_{0}^{L} \nabla p \cdot A\mathbf{w} \, dx dy = 0 . \tag{4.148}$$

It therefore follows that

$$
\begin{aligned}
\dot{J}(\mathbf{w}) &= -\frac{2}{Re} \|A\mathbf{w}\|^2 + 2 \int_{-1}^{1} \int_{0}^{L} ((uu_x + vu_y) Au + (uv_x + vv_y) Av) \, dx dy \\
&\quad + 2 \int_{-1}^{1} \int_{0}^{L} \left(\left(\tilde{U} u_x + \tilde{U}' v \right) Au + \tilde{U} v_x Av \right) dx dy .
\end{aligned}
\tag{4.149}
$$

By Lemma 4.4, Young's inequality and Lemma 4.3, we deduce that (the following c's denoting various positive constants that may vary from line to line and ε being a positive constant that will be chosen small enough later)

$$
\begin{aligned}
\int_{-1}^{1} \int_{0}^{L} uu_x Au \, dx dy &\leq \|u\|_{L^4} \|u_x\|_{L^4} \|Au\| \\
&\leq c \left(\|\nabla u\|^{1/2} \|u\|^{1/2} + \|u\| \right) \left(\|\nabla u_x\|^{1/2} \|u_x\|^{1/2} + \|u_x\| \right) \|Au\| \\
&\leq c\alpha_1 (E, J) + \varepsilon \|A\mathbf{w}\|^2 ,
\end{aligned}
\tag{4.150}
$$

where

$$\alpha_1 (E, J) = E(\mathbf{w}) J(\mathbf{w}) + E^2(\mathbf{w}) J(\mathbf{w}) + J^{3/2}(\mathbf{w}) E^{1/2}(\mathbf{w}) + J^2(\mathbf{w}) E(\mathbf{w}) . \tag{4.151}$$

In the same way, we can estimate other integrals and obtain

$$\int_{-1}^{1} \int_{0}^{L} ((uu_x + vu_y) Au + (uv_x + vv_y) Av) \, dx dy \leq c\alpha_1 (E, J) + \varepsilon \|A\mathbf{w}\|^2 . \tag{4.152}$$

Further we have

$$\int_{-1}^{1} \int_{0}^{L} \left(\left(\tilde{U} u_x + \tilde{U}' v \right) Au + \tilde{U} v_x Av \right) dx dy$$
$$\leq c(\varepsilon) (J(\mathbf{w}) + E(\mathbf{w})) + \varepsilon \|A\mathbf{w}\|^2 . \tag{4.153}$$

Taking ε small enough, we deduce that.

$$\dot{J}(\mathbf{w}) \leq c (E(\mathbf{w}) + J(\mathbf{w}) + \alpha_1 (E, J)) - \frac{1}{Re} \|A\mathbf{w}\|^2 . \tag{4.154}$$

Hence, using (4.143) and applying Lemma 4.1 of [101] with

$$g = c\left(EJ + J^{1/2}E^{1/2}\right), \qquad h = c\left(J + E + EJ + E^2 J\right), \qquad y = J,$$

$$(4.155)$$

and

$$C_1 = c\left(E\left(\mathbf{w}_0\right) + E^2\left(\mathbf{w}_0\right)\right),$$
$$C_2 = c\left(E\left(\mathbf{w}_0\right) + E^2\left(\mathbf{w}_0\right) + E^3\left(\mathbf{w}_0\right)\right), \quad C_3 = cE\left(\mathbf{w}_0\right), \qquad (4.156)$$

we deduce that

$$J\left(\mathbf{w}\left(t\right)\right) \leq \beta_1\left(\mathbf{w}_0\right) e^{-\sigma t}, \qquad \forall t \geq 0, \qquad (4.157)$$

where

$$\beta_1\left(\mathbf{w}_0\right) = c\left(E\left(\mathbf{w}_0\right) + E^2\left(\mathbf{w}_0\right) + E^3\left(\mathbf{w}_0\right)\right.$$
$$\left. + J\left(\mathbf{w}_0\right)\exp\left(c\left(E\left(\mathbf{w}_0\right) + E^2\left(\mathbf{w}_0\right)\right)\right)\right). \quad (4.158)$$

Since $\tau^i \leq ce^\tau$, $e^\tau \leq ee^{\tau^2}$ for $\tau \geq 0$ and $i = 0, 1, 2, 3$ and $E\left(\mathbf{w}_0\right) \leq c\|\mathbf{w}_0\|_{\tilde{\mathbf{H}}^1}^2$, we have

$$\beta_1\left(\mathbf{w}_0\right) \leq c\|\mathbf{w}_0\|_{\tilde{\mathbf{H}}^1}^2 \exp\left(c\|\mathbf{w}_0\|_{\tilde{\mathbf{H}}^1}^4\right). \qquad (4.159)$$

Hence, by Lemma 4.2 and (4.157), we deduce (4.96).

Part 3. We differentiate the first equation of (4.74) with respect to t and multiply it by u_t and integrate over Ω. This gives

$$\int_{-1}^1 \int_0^L u_{tt} u_t \, dx dy = \frac{1}{Re} \int_{-1}^1 \int_0^L u_t \Delta u_t \, dx dy$$
$$- \int_{-1}^1 \int_0^L \left(u_t u_x u_t + u u_{xt} u_t + v_t u_y u_t + v u_{yt} u_t\right) dx dy$$
$$- \int_{-1}^1 \int_0^L \left(\tilde{U} u_{xt} u_t + \tilde{U}' v_t u_t + p_{xt} u_t\right) dx dy. \quad (4.160)$$

Since

$$\int_{-1}^1 \int_0^L u_t \Delta u_t \, dx dy = \int_0^L u_{yt} u_t \big|_{y=-1}^1 \, dx - \int_{-1}^1 \int_0^L \left(u_{xt}^2 + u_{yt}^2\right) dx dy$$
$$= -\frac{1}{k_u} \int_0^L \left(u_t^2\left(x, -1, t\right) + u_t^2\left(x, 1, t\right)\right) dx - \int_{-1}^1 \int_0^L \left(u_{xt}^2 + u_{yt}^2\right) dx dy \quad (4.161)$$

$$\int_{-1}^{1} \int_{0}^{L} (uu_{xt}u_t + vu_{yt}u_t)\, dxdy = \frac{1}{2} \int_{-1}^{1} uu_t^2\big|_{x=0}^{L}\, dy + \frac{1}{2} \int_{0}^{L} vu_t^2\big|_{y=-1}^{1}\, dx$$

$$-\frac{1}{2} \int_{-1}^{1} \int_{0}^{L} (u_x + v_y)\, u_t^2\, dxdy = 0\,, \tag{4.162}$$

$$\int_{-1}^{1} \int_{0}^{L} u_x u_t^2\, dxdy = \int_{-1}^{1} uu_t^2\big|_{x=0}^{L}\, dy - 2 \int_{-1}^{1} \int_{0}^{L} uu_{xt}u_t\, dxdy$$

$$= -2 \int_{-1}^{1} \int_{0}^{L} uu_{xt}u_t\, dxdy\,, \tag{4.163}$$

$$\int_{-1}^{1} \int_{0}^{L} v_t u_y u_t\, dxdy = \int_{-1}^{1} v_t uu_t\big|_{y=-1}^{1}\, dy - \int_{-1}^{1} \int_{0}^{L} (uv_{yt}u_t + uv_t u_{yt})\, dxdy$$

$$= \int_{-1}^{1} \int_{0}^{L} (uu_{xt}u_t - uv_t u_{yt})\, dxdy\,, \tag{4.164}$$

$$\int_{-1}^{1} \int_{0}^{L} \tilde{U} u_{xt}u_t\, dxdy = \frac{1}{2} \int_{-1}^{1} \tilde{U} u_t^2\big|_{x=0}^{L}\, dy = 0\,, \tag{4.165}$$

$$\int_{-1}^{1} \int_{0}^{L} p_{xt}u_t\, dxdy = \int_{-1}^{1} p_t u_t\big|_{x=0}^{L}\, dy - \int_{-1}^{1} \int_{0}^{L} p_t u_{xt}\, dxdy$$

$$= -\int_{-1}^{1} \int_{0}^{L} p_t u_{xt}\, dxdy\,, \tag{4.166}$$

we deduce that

$$\frac{1}{2} \frac{d}{dt} \left(\int_{-1}^{1} \int_{0}^{L} u_t^2\, dxdy \right) = -\frac{1}{k_u Re} \int_{0}^{L} \left(u_t^2(x,-1,t) + u_t^2(x,1,t) \right)\, dx$$

$$-\frac{1}{Re} \int_{-1}^{1} \int_{0}^{L} \left(u_{xt}^2 + u_{yt}^2 \right)\, dxdy + \int_{-1}^{1} \int_{0}^{L} (uu_{xt}u_t + uv_t u_{yt})\, dxdy$$

$$-\int_{-1}^{1} \int_{0}^{L} \tilde{U}' v_t u_t\, dxdy + \int_{-1}^{1} \int_{0}^{L} p_t u_{xt}\, dxdy\,. \tag{4.167}$$

Differentiating the second equation of (4.74) with respect to t, multiplying it by v_t and integrating over Ω, we obtain

$$\int_{-1}^{1} \int_{0}^{L} v_{tt}v_t\, dxdy = \frac{1}{Re} \int_{-1}^{1} \int_{0}^{L} v_t \Delta v_t\, dxdy$$

$$-\int_{-1}^{1} \int_{0}^{L} (u_t v_x v_t + uv_{xt}v_t + v_t v_y v_t + vv_{yt}v_t)\, dxdy$$

$$-\int_{-1}^{1} \int_{0}^{L} \left(\tilde{U} v_{xt}v_t + p_{yt}v_t \right)\, dxdy\,. \tag{4.168}$$

Since

$$\int_{-1}^{1} \int_0^L v_t \Delta v_t \, dx dy = \int_0^L v_{yt} v_t \big|_{y=-1}^{1} \, dx - \int_{-1}^{1} \int_0^L \left(v_{xt}^2 + v_{yt}^2 \right) \, dx dy$$
$$= -\int_{-1}^{1} \int_0^L \left(v_{xt}^2 + v_{yt}^2 \right) \, dx dy, \qquad (4.169)$$

$$\int_{-1}^{1} \int_0^L \left(u v_{xt} v_t + v v_{yt} v_t \right) \, dx dy = \frac{1}{2} \int_{-1}^{1} u v_t^2 \big|_{x=0}^{L} \, dy + \frac{1}{2} \int_0^L v v_t^2 \big|_{y=-1}^{1} \, dx$$
$$- \frac{1}{2} \int_{-1}^{1} \int_0^L \left(u_x + v_y \right) v_t^2 \, dx dy = 0, \qquad (4.170)$$

$$\int_{-1}^{1} \int_0^L v_y v_t^2 \, dx dy = \int_0^L v v_t^2 \big|_{y=-1}^{1} \, dx - 2 \int_{-1}^{1} \int_0^L v v_{yt} v_t \, dx dy$$
$$= -2 \int_{-1}^{1} \int_0^L v v_{yt} v_t \, dx dy, \qquad (4.171)$$

$$\int_{-1}^{1} \int_0^L u_t v_x v_t \, dx dy = \int_{-1}^{1} u_t v v_t \big|_{x=0}^{L} \, dy - \int_{-1}^{1} \int_0^L \left(v u_{xt} v_t + v u_t v_{xt} \right) \, dx dy$$
$$= \int_{-1}^{1} \int_0^L \left(v v_{yt} v_t - v u_t v_{xt} \right) \, dx dy, \qquad (4.172)$$

$$\int_{-1}^{1} \int_0^L \tilde{U} v_{xt} v_t \, dx dy = \frac{1}{2} \int_{-1}^{1} \tilde{U} v_t^2 \big|_{x=0}^{L} \, dy = 0, \qquad (4.173)$$

$$\int_{-1}^{1} \int_0^L p_{yt} v_t \, dx dy = \int_0^L p_t v_t \big|_{y=-1}^{1} \, dx - \int_{-1}^{1} \int_0^L p_t v_{yt} \, dx dy$$
$$= -\int_{-1}^{1} \int_0^L p_t v_{yt} \, dx dy, \qquad (4.174)$$

we deduce that

$$\frac{1}{2} \frac{d}{dt} \left(\int_{-1}^{1} \int_0^L v_t^2 \, dx dy \right) = -\frac{1}{Re} \int_{-1}^{1} \int_0^L \left(v_{xt}^2 + v_{yt}^2 \right) \, dx dy$$
$$+ \int_{-1}^{1} \int_0^L \left(v v_{yt} v_t + v u_t v_{xt} \right) \, dx dy + \int_{-1}^{1} \int_0^L p_t v_{yt} \, dx dy. \qquad (4.175)$$

It therefore follows from (4.118), (4.167) and (4.175) that

$$
\begin{aligned}
\dot{E}\left(\mathbf{w}_t\right) = \ & -\frac{2}{Re} \int_{-1}^{1} \int_{0}^{L} \left(u_{xt}^2 + u_{yt}^2 + v_{xt}^2 + v_{yt}^2\right) dx dy \\
& -\frac{2}{k_u Re} \int_{0}^{L} \left(u_t^2\left(x, -1, t\right) + u_t^2\left(x, 1, t\right)\right) dx \\
& +2 \int_{-1}^{1} \int_{0}^{L} \left(u u_{xt} u_t + u v_t u_{yt} + v v_{yt} v_t + v u_t v_{xt}\right) dx dy \\
& -2 \int_{-1}^{1} \int_{0}^{L} \tilde{U}' v_t u_t \, dx dy \\
\leq \ & -\frac{1}{Re} J\left(\mathbf{w}_t\right) + 2\left(1 - \frac{1}{8Re}\right) E\left(\mathbf{w}_t\right) + \frac{1}{Re} \int_{0}^{L} u_t^2\left(x, -1, t\right) dx \\
& -\frac{1}{k_u Re} \int_{0}^{L} \left(u_t^2\left(x, -1, t\right) + u_t^2\left(x, 1, t\right)\right) dx \\
& +2 \int_{-1}^{1} \int_{0}^{L} \left(u u_{xt} u_t + u v_t u_{yt} + v v_{yt} v_t + v u_t v_{xt}\right) dx dy. \quad (4.176)
\end{aligned}
$$

By Lemma 4.4 and Young's inequality, we deduce that (the following c's denoting various positive constants that may vary from line to line and ε being a positive constant that will be determined later)

$$
\begin{aligned}
\int_{-1}^{1} \int_{0}^{L} u u_{xt} u_t \, dx dy \ & \leq \ \|u\|_{L^4} \|u_t\|_{L^4} \|u_{xt}\| \\
& \leq \ c\left(\|\nabla u\|^{1/2} \|u\|^{1/2} + \|u\|\right) \\
& \quad \times \left(\|\nabla u_t\|^{1/2} \|u_t\|^{1/2} + \|u_t\|\right) \|u_{xt}\| \\
& \leq \ c\left(J^{1/4}\left(\mathbf{w}\right) E^{1/4}\left(\mathbf{w}\right) + E^{1/2}\left(\mathbf{w}\right)\right) \\
& \quad \times \left(J^{3/4}\left(\mathbf{w}_t\right) E^{1/4}\left(\mathbf{w}_t\right) + E^{1/2}\left(\mathbf{w}_t\right) J^{1/2}\left(\mathbf{w}_t\right)\right) \\
& \leq \ c\alpha_2\left(E, J\right) E\left(\mathbf{w}_t\right) + \varepsilon J\left(\mathbf{w}_t\right), \quad (4.177)
\end{aligned}
$$

where

$$
\alpha_2\left(E, J\right) = E\left(\mathbf{w}\right) J\left(\mathbf{w}\right) + E^2\left(\mathbf{w}\right) + J\left(\mathbf{w}\right) + E\left(\mathbf{w}\right). \quad (4.178)
$$

Similarly, we have

$$
\int_{-1}^{1} \int_{0}^{L} u u_{yt} v_t \, dx dy \ \leq \ c\alpha_2\left(E, J\right) E\left(\mathbf{w}_t\right) + \varepsilon J\left(\mathbf{w}_t\right), \quad (4.179)
$$

$$
\int_{-1}^{1} \int_{0}^{L} v v_{yt} v_t \, dx dy \ \leq \ c\alpha_2\left(E, J\right) E\left(\mathbf{w}_t\right) + \varepsilon J\left(\mathbf{w}_t\right), \quad (4.180)
$$

$$\int_{-1}^{1}\int_{0}^{L} vv_{xt}u_t\, dxdy \;\leq\; c\alpha_2\left(E,J\right)E\left(\mathbf{w}_t\right)+\varepsilon J\left(\mathbf{w}_t\right). \qquad (4.181)$$

It therefore follows from (4.176) that

$$\dot{E}\left(\mathbf{w}_t\right)\leq\left(\varepsilon-\frac{1}{Re}\right)J\left(\mathbf{w}_t\right)-\sigma E\left(\mathbf{w}_t\right)+c\alpha_2\left(E,J\right)E\left(\mathbf{w}_t\right) \qquad (4.182)$$

which implies

$$\frac{d}{dt}\left(e^{\sigma t}E\left(\mathbf{w}_t\right)\right)\leq c\alpha_2\left(E,J\right)e^{\sigma t}E\left(\mathbf{w}_t\right), \qquad (4.183)$$

where σ is given by (4.94). Therefore, by (4.143) and Gronwall's inequality (see. e.g., [86, p.63]), we deduce that

$$E\left(\mathbf{w}_t\left(t\right)\right)\leq E\left(\mathbf{w}_t\left(0\right)\right)\exp\left(cE\left(\mathbf{w}_0\right)\left(E\left(\mathbf{w}_0\right)+1\right)\right)e^{-\sigma t}, \qquad \forall t\geq 0. \quad (4.184)$$

On the other hand, by (4.144), (4.147) and (4.148), we have

$$\frac{1}{Re}\left\|A\mathbf{w}\right\|^2 = \int_{-1}^{1}\int_{0}^{L}\left(\mathbf{w}_t\cdot A\mathbf{w}+\left(uu_x+vu_y\right)Au+\left(uv_x+vv_y\right)Av\right)dxdy$$

$$+\int_{-1}^{1}\int_{0}^{L}\left(\left(\tilde{U}u_x+\tilde{U}'v\right)Au+\tilde{U}v_x Av\right)dxdy. \qquad (4.185)$$

Using (4.152) and (4.153) we obtain

$$\frac{1}{Re}\left\|A\mathbf{w}\right\|^2\leq c\left(E\left(\mathbf{w}_t\right)+\alpha_3\left(E,J\right)\right)+\varepsilon\left\|A\mathbf{w}\right\|^2, \qquad (4.186)$$

where

$$\alpha_3\left(E,J\right)=E\left(\mathbf{w}\right)+J\left(\mathbf{w}\right)+\alpha_1\left(E,J\right). \qquad (4.187)$$

Hence, by (4.95), (4.157) and (4.184), we deduce that

$$\left\|A\mathbf{w}\right\|^2\leq\beta_2\left(\mathbf{w}_0\right)e^{-\sigma t}, \qquad \forall t\geq 0, \qquad (4.188)$$

where

$$\beta_2\left(\mathbf{w}_0\right)=c\left(E\left(\mathbf{w}_t\left(0\right)\right)+\sum_{i=1}^{7}E^i\left(\mathbf{w}_0\right)+\sum_{i=1}^{4}J^i\left(\mathbf{w}_0\right)\right)$$

$$\times\exp\left(cE\left(\mathbf{w}\left(0\right)\right)\left(E\left(\mathbf{w}\left(0\right)\right)+1\right)\right). \qquad (4.189)$$

In addition, multiplying (4.74) by $\mathbf{w}_t$, as in the proof of (4.186), we can prove that

$$E\left(\mathbf{w}_t\right)\leq c\left(\left\|\Delta\mathbf{w}\right\|^2+\alpha_3\left(E,J\right)\right)+\varepsilon E\left(\mathbf{w}_t\right), \qquad (4.190)$$

which implies that

$$E\left(\mathbf{w}_t\left(0\right)\right) \le c\left(\|\mathbf{w}_0\|_{\tilde{\mathbf{H}}^2}^2 + \sum_{i=1}^{4}\left(E^i\left(\mathbf{w}_0\right) + J^i\left(\mathbf{w}_0\right)\right)\right). \tag{4.191}$$

Thus, as in (4.159), we deduce that

$$
\begin{aligned}
\beta_2\left(\mathbf{w}_0\right) &\le c\left(\|\mathbf{w}_0\|_{\tilde{\mathbf{H}}^2}^2 + \sum_{i=1}^{7} E^i\left(\mathbf{w}_0\right) + \sum_{i=1}^{4} J^i\left(\mathbf{w}_0\right)\right) \\
&\qquad \times \exp\left(cE\left(\mathbf{w}_0\right)\left(E\left(\mathbf{w}_0\right)+1\right)\right). \\
&\le c\|\mathbf{w}_0\|_{\tilde{\mathbf{H}}^2}^2 \exp\left(c\|\mathbf{w}_0\|_{\tilde{\mathbf{H}}^2}^2\right). \tag{4.192}
\end{aligned}
$$

Hence, by (4.188) and Lemma 4.3, we deduce (4.97) and inequalities (4.184) and (4.191) show the stated bound of $\|\mathbf{w}_t\left(t\right)\|$.

Multiplying the first equation of (4.74) by p_x and the second equation of (4.74) by p_y, integrating over Ω and using (4.152) and (4.153) with $A\mathbf{w}$ replaced by ∇p, we obtain

$$
\begin{aligned}
\|\nabla p\left(t\right)\|^2 &= -\int_{-1}^{1}\int_{0}^{L}\left(\mathbf{w}_t\cdot\nabla p - \frac{1}{Re}\Delta\mathbf{w}\cdot\nabla p + \left(uu_x + vu_y\right)p_x\right. \\
&\qquad\qquad\qquad \left. + \left(uv_x + vv_y\right)\right)\,dxdy \\
&\quad - \int_{-1}^{1}\int_{0}^{L}\left(\left(\tilde{U}u_x + \tilde{U}'v\right)p_x + \tilde{U}v_x p_y\right)\,dxdy \\
&\le c\left(E\left(\mathbf{w}_t\right) + \|\Delta\mathbf{w}\|^2 + \alpha_3\left(E,J\right)\right) + \varepsilon\|\nabla p\|^2. \tag{4.193}
\end{aligned}
$$

From this last inequality the stated bound on $\|\nabla p\|$ follows by (4.95), (4.96) and (4.97).

Existence and regularity. We use the Galerkin method to prove existence of solutions. We look for an approximate solution in the form

$$\mathbf{w}^n\left(x,y,t\right) = \sum_{i=1}^{n} c_{in}\left(t\right)\Phi_i\left(x,y\right), \qquad n = 1,2,3,\ldots \tag{4.194}$$

where the set $\{\Phi_i\}_{i\ge 1}$ forms a Riesz basis in $D\left(A\right)$. We require that $\mathbf{w}^n$ satisfies (4.92), i.e.

$$
\begin{aligned}
\int_{-1}^{1}\int_{0}^{L}&\left(\mathbf{w}_t^n\cdot\Phi_i + \frac{1}{Re}\mathrm{Tr}\left\{\nabla\mathbf{w}^n\nabla\Phi_i\right\}\right. \\
&\quad \left. + \left(\mathbf{w}^n\cdot\nabla\right)\mathbf{w}^n\cdot\Phi_i + \tilde{U}\mathbf{w}_x^n\Phi_i + \tilde{U}'v^n\xi_i\right)\,dxdy \\
&= -\frac{1}{k_u Re}\int_{0}^{L}\left(u^n\left(x,1,t\right)\xi_i\left(x,1\right) + u^n\left(x,-1,t\right)\xi_i\left(x,-1\right)\right)\,dx \tag{4.195}
\end{aligned}
$$

for all $\boldsymbol{\Phi}_i = (\xi_i, \eta_i)$, $i = 1, \ldots, n$. Expanding the definition of $\mathbf{w}^n$, equation (4.195) provides us with a system of first order ordinary differential equations for the time dependent coefficients $\{c_{in}(t)\}_{i \geq 1}$, where we choose the set of initial conditions

$$c_{in}(0) = \int_{-1}^{1} \int_{0}^{L} \mathbf{w}_0(x, y) \cdot \boldsymbol{\Phi}_i(x, y) \, dx dy \qquad i = 1, \ldots, n. \qquad (4.196)$$

This system depends on $\{c_{in}\}_{i \geq 1}$ analytically, hence, in order to show the existence of a unique solution for all $t \in [0, T]$, it is sufficient to verify the boundedness of $\{|c_{in}(t)|\}_{i \geq 1}$. This is equivalent to the boundedness of the norms $\{\|\mathbf{w}^n(t)\|\}_{n \geq 1}$ as a consequence of the system $\{\boldsymbol{\Phi}_i\}_{i \geq 1}$ being a Riesz basis. Replacing $\boldsymbol{\Phi}_i$ by $\mathbf{w}^n$ in (4.195) we deduce estimates (4.95) and (4.143) for $\mathbf{w}^n$. Namely

$$\|\mathbf{w}^n(t)\| \leq \|\mathbf{w}_0^n\| e^{-\sigma t} \leq \|\mathbf{w}_0\| e^{-\sigma t}, \qquad (4.197)$$

and

$$\int_0^T e^{\sigma t} \|\mathbf{w}^n(t)\|_{\tilde{\mathbf{V}}}^2 \, dt \leq M \|\mathbf{w}_0^n\| \leq M \|\mathbf{w}_0\| \qquad (4.198)$$

for some constants M and σ and for all $t \in [0, T]$. In these calculations the steps are justified using the regularity of $\mathbf{w}^n$.

The next step in Galerkin's method is to show that a subsequence of approximating solutions $\{\mathbf{w}^n\}_{n \geq 1}$ converges to a limiting function $\mathbf{w}$ as $n \to \infty$. The convergence is obtained using compactness arguments. In our case, by the uniform boundedness of the sequence $\{\mathbf{w}^n\}_{n \geq 1}$ in $L^2\left([0, T]; \tilde{\mathbf{V}}\right) \cap L^\infty\left([0, T]; \tilde{\mathbf{H}}\right)$ a subsequence $\{\mathbf{w}^n\}_{n \geq 1}$ converges to some element $\mathbf{w} \in L^2\left([0, T]; \tilde{\mathbf{V}}\right) \cap L^\infty\left([0, T]; \tilde{\mathbf{H}}\right)$. The convergence is weak in $L^2\left([0, T]; \tilde{\mathbf{V}}\right)$, weak–star in $L^\infty\left([0, T]; \tilde{\mathbf{H}}\right)$ and, due to compactness ([127, pp. 285–287],) strong in $L^2\left([0, T]; \tilde{\mathbf{V}}\right)$. These convergence properties enable us to prove, as a final step of Galerkin's method, that the limiting function $\mathbf{w}$ is in fact a weak solution of (4.92). We have to show that each term of equation

$$\frac{d}{dt} \int_{-1}^{1} \int_{0}^{L} \mathbf{w}^n \cdot \boldsymbol{\Phi} \, dx dy + \frac{1}{Re} \int_{-1}^{1} \int_{0}^{L} \mathrm{Tr}\left\{ \nabla \mathbf{w}^{nT} \nabla \boldsymbol{\Phi} \right\} \, dx dy$$

$$+ \int_{-1}^{1} \int_{0}^{L} (\mathbf{w}^n \cdot \nabla) \mathbf{w}^n \cdot \boldsymbol{\Phi} \, dx dy + \int_{-1}^{1} \int_{0}^{L} \tilde{U} \mathbf{w}_x^n \boldsymbol{\Phi} \, dx dy + \int_{-1}^{1} \int_{0}^{L} \tilde{U}' v^n \xi \, dx dy$$

$$= -\frac{1}{k_u Re} \int_{0}^{L} (u^n(x, 1, t) \xi(x, 1) + u^n(x, -1, t) \xi(x, -1)) dx \qquad (4.199)$$

converges to the corresponding term of equation

$$
\frac{d}{dt} \int_{-1}^{1} \int_{0}^{L} \mathbf{w} \cdot \mathbf{\Phi} \, dx \, dy + \frac{1}{Re} \int_{-1}^{1} \int_{0}^{L} \mathrm{Tr}\left\{ \nabla \mathbf{w}^T \nabla \mathbf{\Phi} \right\} \, dx \, dy
$$

$$
+ \int_{-1}^{1} \int_{0}^{L} (\mathbf{w} \cdot \nabla) \mathbf{w} \cdot \mathbf{\Phi} \, dx \, dy + \int_{-1}^{1} \int_{0}^{L} \tilde{U} \mathbf{w}_x \mathbf{\Phi} \, dx \, dy + \int_{-1}^{1} \int_{0}^{L} \tilde{U}' v \xi \, dx \, dy
$$

$$
= -\frac{1}{k_u Re} \int_{0}^{L} (u(x,1,t)\xi(x,1) + u(x,-1,t)\xi(x,-1)) \, dx \quad (4.200)
$$

for all $\mathbf{\Phi} = (\xi, \eta) \in \tilde{\mathbf{V}}$. This is a standard step in the theory of Navier–Stokes equations for all the terms except the ones on the right hand side of equations (4.199) and (4.200). These terms are present due to our special boundary conditions (4.77). We prove here the convergence of the first term on the right. The convergence of the second term can be proved in the same way. We have to show that

$$
\int_{0}^{L} u^n(x,1,t)\xi(x,1) \, dx \xrightarrow{n\to\infty} \int_{0}^{L} u(x,1,t)\xi(x,1) \, dx \quad (4.201)
$$

for all $\mathbf{\Phi} = (\xi, \eta) \in \tilde{\mathbf{V}}$. We take the difference of the two sides in (4.201) and take the $L^2[0,T]$–inner product of the result by a function $c(t) \in L^2(0,T)$. We obtain

$$
\int_{0}^{T} \left(\int_{0}^{L} u^n(x,1,t)c(t)\xi(x,1) \, dx - \int_{0}^{L} u(x,1,t)c(t)\xi(x,1) \, dx \right) dt
$$

$$
\leq \|\xi\|_{L^\infty} \int_{0}^{T} c(t) \int_{0}^{L} \sup_{y\in(-1,1)} |u^n(x,y,t) - u(x,y,t)| \, dx \, dt
$$

$$
\leq M \|\xi\|_{L^\infty} \int_{0}^{T} c(t) \int_{0}^{L} \left(\int_{-1}^{1} |u^n - u|^2 \, dy \right)^{1/2} dx \, dt
$$

$$
+ M \|\xi\|_{L^\infty} \int_{0}^{T} c(t) \int_{0}^{L} \left(\int_{-1}^{1} |u_y^n - u_y|^2 \, dy \right)^{1/4}
$$

$$
\times \left(\int_{-1}^{1} |u^n - u|^2 \, dy \right)^{1/4} dx \, dt , \quad (4.202)
$$

where we used the one–dimensional equivalent of inequality (4.127). We further estimate expressions from (4.202)

$$
\int_{0}^{T} c(t) \int_{0}^{L} \left(\int_{-1}^{1} |u^n - u|^2 \, dy \right)^{1/2} dx \, dt
$$

$$
\leq \int_{0}^{T} c(t) \left(\int_{-1}^{1} \int_{0}^{L} |u^n - u|^2 \, dx \, dy \right)^{1/2} dt
$$

$$\leq \left(\int_0^T c^2(t)\, dt \right)^{1/2} \left(\int_0^T \|\mathbf{w}^n - \mathbf{w}\|^2\, dt \right)^{1/2}. \qquad (4.203)$$

Here $\int_0^T \|\mathbf{w}^n - \mathbf{w}\|^2\, dt$ converges to zero as $n \to \infty$ according to the strong convergence in $L^2\left([0,T];\tilde{\mathbf{H}}\right)$. The last expression in (4.202) can be estimated the following way:

$$\int_0^T c(t) \int_0^L \left(\int_{-1}^1 |u_y^n - u_y|^2\, dy \right)^{1/4} \left(\int_{-1}^1 |u^n - u|^2\, dy \right)^{1/4} dx\, dt$$

$$\leq \int_0^T c(t) \|\nabla(\mathbf{w}^n - \mathbf{w})\|^{1/2} \|\mathbf{w}^n - \mathbf{w}\|^{1/2}\, dt$$

$$\leq \sup_{t \in [0,T]} \left(\|\mathbf{w}^n\|_{\tilde{\mathbf{V}}} + \|\mathbf{w}\|_{\tilde{\mathbf{V}}} \right)^{1/2} \left(\int_0^T c^2(t)\, dt \right)^{1/2} \left(\int_0^T \|\mathbf{w}^n - \mathbf{w}\|\, dt \right)^{1/2}$$

$$\leq \sup_{t \in [0,T]} \left(\|\mathbf{w}^n\|_{\tilde{\mathbf{V}}} + \|\mathbf{w}\|_{\tilde{\mathbf{V}}} \right)^{1/2}$$

$$\times \left(\int_0^T c^2(t)\, dt \right)^{1/2} \sqrt{T} \left(\int_0^T \|\mathbf{w}^n - \mathbf{w}\|^2\, dt \right)^{1/4}. \qquad (4.204)$$

Here the last factor converges to zero while the other factors are bounded as $n \to \infty$. Since $c(t) \in L^2(0,T)$ was arbitrary, we obtain the desired convergence result.

It follows from the Helmholtz decomposition (4.101)–(4.102) that, once the existence of weak solutions $\mathbf{w}$ is established, we obtain the existence of pressure p, so that (4.74)–(4.77) are satisfied in a distributional sense.

The rest of the regularity statements in Theorem 4.1 follows from estimates (4.157), (4.96), (4.184), (4.188), (4.97) and from embedding theorems.

Continuous dependence on initial data and uniqueness. Let $\mathbf{w}_1 = (u_1, v_1)^T$, and $\mathbf{w}_2 = (u_2, v_2)^T$, p_2 be two solutions of (4.74)–(4.77) corresponding to initial data $\mathbf{w}_1^0$ and $\mathbf{w}_2^0$ respectively. Their difference $\mathbf{w} = (u, v)^T = \mathbf{w}_1 - \mathbf{w}_2$, $p = p_1 - p_2$ satisfies

$$u_t - \frac{1}{Re}\Delta u + u_1 u_x + u u_{2x} + v_1 u_y + v u_{2y} + \tilde{U} u_x + \tilde{U}' v + p_x = 0, \qquad (4.205)$$

$$v_t - \frac{1}{Re}\Delta v + u_1 v_x + u v_{2x} + v_1 v_y + v v_{2y} + \tilde{U} v_x + p_y = 0, \qquad (4.206)$$

$$u_x + v_y = 0, \qquad (4.207)$$

with boundary condition (4.75)–(4.77). Taking the scalar product of (4.205) with u we obtain

$$\int_{-1}^{1}\int_{0}^{L} u_t u \, dx dy - \frac{1}{Re}\int_{-1}^{1}\int_{0}^{L}\Delta u u \, dx dy + \int_{-1}^{1}\int_{0}^{L} u_1 u_x u \, dx dy$$

$$+ \int_{-1}^{1}\int_{0}^{L} u u_{2x} u \, dx dy + \int_{-1}^{1}\int_{0}^{L} v_1 u_y u \, dx dy + \int_{-1}^{1}\int_{0}^{L} v u_{2y} u \, dx dy$$

$$+ \int_{-1}^{1}\int_{0}^{L}\tilde{U} u_x u \, dx dy + \int_{-1}^{1}\int_{0}^{L}\tilde{U}' v u \, dx dy + \int_{-1}^{1}\int_{0}^{L} p_x u \, dx dy = 0 . \quad (4.208)$$

Here

$$\int_{-1}^{1}\int_{0}^{L} u_1 u_x u \, dx dy = \frac{1}{2}\int_{-1}^{1} u_1 u^2\big|_{x=0}^{L} \, dy - \frac{1}{2}\int_{-1}^{1}\int_{0}^{L} u_{1x} u^2 \, dx dy$$

$$\leq \frac{1}{2}\|\nabla \mathbf{w}_1\| \|\mathbf{w}\|_{L^4}^2$$

$$\leq M \|\nabla \mathbf{w}_1\| \left(\|\mathbf{w}\| + \|\nabla \mathbf{w}\|^{1/2}\|\mathbf{w}\|^{1/2}\right)^2$$

$$= M \|\nabla \mathbf{w}_1\| \|\mathbf{w}\|^2 + M \|\nabla \mathbf{w}\|^{1/2}\|\nabla \mathbf{w}_1\| \|\mathbf{w}\|^{3/2}$$

$$+ M \|\nabla \mathbf{w}\| \|\nabla \mathbf{w}_1\| \|\mathbf{w}\|$$

$$\leq M \|\nabla \mathbf{w}_1\| \|\mathbf{w}\|^2 + \frac{\delta}{2}\|\nabla \mathbf{w}\|^2 + M \|\nabla \mathbf{w}_1\|^{4/3}\|\mathbf{w}\|^2$$

$$+ \frac{\delta}{2}\|\nabla \mathbf{w}\|^2 + M \|\nabla \mathbf{w}_1\|^2 \|\mathbf{w}\|^2$$

$$\leq \delta \|\nabla \mathbf{w}\|^2 + M \left(\|\nabla \mathbf{w}_m\|\right)\|\mathbf{w}\|^2 , \quad (4.209)$$

where we used Young's inequality twice in the fourth step with $\delta > 0$ arbitrary and

$$M\left(\|\nabla \mathbf{w}_m(t)\|\right) \equiv c \max_{i=1,2}\left(\|\nabla \mathbf{w}_i\| + \|\nabla \mathbf{w}_i\|^{4/3} + \|\mathbf{w}_i(t)\|^2\right). \quad (4.210)$$

Terms 4, 5 and 6 in (4.208) can be estimated the same way. The rest of the terms are estimated as in obtaining (4.95). Taking the scalar product of (4.206) with v we obtain

$$\int_{-1}^{1}\int_{0}^{L} v_t v \, dx dy - \frac{1}{Re}\int_{-1}^{1}\int_{0}^{L}\Delta v v \, dx dy + \int_{-1}^{1}\int_{0}^{L} u_1 v_x v \, dx dy$$

$$+ \int_{-1}^{1}\int_{0}^{L} u v_{2x} v \, dx dy + \int_{-1}^{1}\int_{0}^{L} v_1 v_y v \, dx dy + \int_{-1}^{1}\int_{0}^{L} v v_{2y} v \, dx dy$$

$$+ \int_{-1}^{1}\int_{0}^{L}\tilde{U} v_x v \, dx dy + \int_{-1}^{1}\int_{0}^{L} p_y v \, dx dy = 0 . \quad (4.211)$$

The estimation of the terms is similar to (4.208). We obtain from (4.208) and (4.211), after choosing appropriate δ,

$$\frac{d}{dt} \left\| \mathbf{w}\left(t\right) \right\|^2 \le M \left(\left\| \nabla \mathbf{w}_m \left(t\right) \right\| \right) \left\| \mathbf{w}\left(t\right) \right\|^2, . \tag{4.212}$$

Gronwall's inequality applied to (4.212) implies that

$$\left\| \mathbf{w}\left(t\right) \right\|^2 \le \left\| \mathbf{w}\left(0\right) \right\|^2 \exp \left(\int_0^T M \left(\left\| \nabla \mathbf{w}_m \left(\tau\right) \right\| \right) d\tau \right) \tag{4.213}$$

for all $t \in [0, T]$. Since $M \left(\left\| \nabla \mathbf{w}_m \left(t\right) \right\| \right)$ is integrable over every finite interval $[0, T]$, (4.213) proves the continuous dependence of solutions on the initial data in the L^2 norm.

Numerical Simulation

The simulation example in this section is performed in a channel of length 4π and height 2 for Reynolds number $Re = 15000$, which is five orders of magnitude greater than required in Theorem 4.1, and is three times the critical value (5772, corresponding to loss of linear stability) for 2D channel flow. The validity of the stabilization result beyond the assumptions of Theorem 4.1 is not completely surprising since our Lyapunov analysis is based on conservative energy estimates.[2] The control gain used is $k_u = 1$.

A hybrid Fourier pseudospectral–finite difference discretization and the fractional step technique based on a hybrid Runge–Kutta/Crank–Nicolson time discretization was used to generate the results. The code originally has been adapted from a Fourier–Chebyshev pseudospectral code of T. Bewley [27], changing the wall-normal discretization to second–order finite differences (P. Blossey, private communication). The nonlinear terms in the Navier–Stokes equations are integrated explicitly using a fourth–order, low storage Runge–Kutta method first devised by Carpenter and Kennedy [36]. The viscous terms are treated implicitly using the Crank–Nicolson method. The numerical method uses "constant volume flux per unit span" instead of the "constant average pressure gradient" assumption to speed up computations. The differences between the two cases are discussed in, for example, [118]. The number of grid points used in our computations was 128×120 and the (adaptive) time step was

---

[2] The effect of boundary control law (4.77) can be seen mathematically in inequality (4.138) in the context of the L^2 perturbation energy. The boundary integral

$$\int_0^L \left(\frac{2}{Re} \left(1 - \frac{1}{k_u} \right) u^2 \left(x, -1, t\right) - \frac{2}{k_u Re} u^2 \left(x, 1, t\right) \right) dx \tag{4.214}$$

is negative even for large Reynolds numbers if k_u is sufficiently small. Hence, it improves the stability properties in general. The trace theorem however does not allow us to compare this term and the total energy and to prove the stability results of Theorem 4.1 for large Reynolds numbers. This shows the need for numerical simulation.

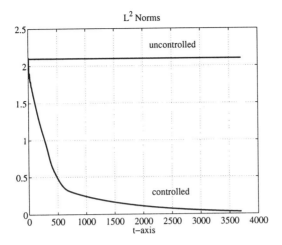

Figure 4.15: Energy comparison [16].

in the range of $0.05 - 0.07$. The grid points had hyperbolic tangent $(y_j = 1 + \tanh\left(s\left(2\frac{j}{NY} - 1\right)\right) / \tanh(s) \quad j = 0, \ldots, NY)$ distribution with stretching factor $s = 1.75$ in the vertical direction in order to achieve high resolution in the critical boundary layer. In order to obtain the flow at the walls in the controlled case the quadratic Three–Point Endpoint Formula was used to approximate the derivatives at the boundary $(U_y(x, 0, t), U_y(x, 2, t))$. This formula is applied in a semi-implicit way in order to avoid numerical instabilities. Namely, the Three–Point Endpoint Formula at the bottom wall has the form

$$U_y(-1) \approx d_0 U_0 + d_1 U_1 + d_2 U_2, \tag{4.215}$$

with notation $U_j = U(y_j)$, $j = 0, 1, 2$ and with appropriate constants d_0, d_1 and d_2. We can write control law (4.66) now as

$$U_0^{n+1} = k_u \left[d_0 U_0^{n+1} + d_1 U_1^n + d_2 U_2^n - 2\right], \tag{4.216}$$

where superscripts n and $n + 1$ refer to values at time step n and $n + 1$ respectively. Equation (4.216) results in the update law

$$U_0^{n+1} = k_u \left(d_1 U_1^n + d_2 U_2^n - 2\right) / (1 - k_u d_0) \tag{4.217}$$

at the boundary. The boundary condition at the top wall is updated in a similar way. The numerical results show very good agreement with results obtained from a finite volume code used at early stages of simulations. As initial data we consider a statistically steady state flow field obtained from a random perturbation of the parabolic profile over a large time period using the uncontrolled system.

Figure 4.15 shows that our controller achieves stabilization. This is expressed in terms of the L^2–norm of the error between the steady state and the actual

Uncontrolled

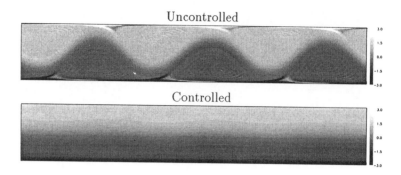

Controlled

Figure 4.16: Vorticity maps at $t = 700$ [16].

Uncontrolled

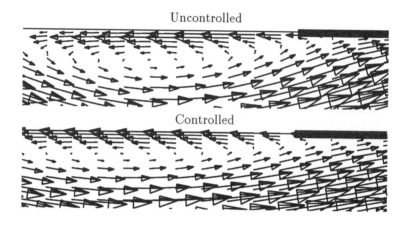

Controlled

Figure 4.17: Recirculation in the flow at $t = 120$, in a rectangle of dimension 1.37×0.31 zoomed out of a channel of dimension $4\pi \times 2$. The shaded region (upper right corner) is magnified in Figure 4.18 [16].

velocity field, the so called perturbation energy, which corresponds to system (4.74)–(4.77) with $k_u = 0$ (zero Dirichlet boundary conditions on the walls) in the uncontrolled case. The initially fast perturbation energy decay somewhat slows down for larger time. What we see here is an interesting example of interaction between linear and nonlinear behavior in a dynamical system. Initially, when the velocity perturbations are large, and the flow is highly nonlinear (exhibiting Tollmien–Schlichting waves with recirculation, see the uncontrolled flow in Figures 4.16 and 4.17). The strong convective (quadratic) nonlinearity dominates over the linear dynamics and the energy decay is fast. Later, at about $t = 500$, the recirculation disappears, the controlled flow becomes close to laminar, and linear behavior dominates, along with its exponential energy decay (with small decay rate).

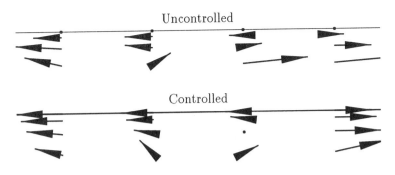

Figure 4.18: Velocity field in a rectangle of dimension 0.393×0.012 zoomed out of a channel of dimension $4\pi \times 2$, at time $t = 120$. The control (thick arrows) acts both *downstream* and *upstream*. The control maintains the value of shear near the desired (laminar) steady-state value [16].

In the vorticity map, depicted in Figure 4.16 it is striking how uniform the vorticity field becomes for the controlled case, while we observe quasi–periodic bursting (cf. [77]) in the uncontrolled case. We obtained similar vorticity maps of the uncontrolled flow for other (lower) Reynolds numbers, that show agreement qualitatively with the vorticity maps obtained by Jiménez [77]. His paper explains the generation of vortex blobs at the wall along with their ejection into the channel and their final dissipation by viscosity in the uncontrolled case.

The uniformity of the wall shear stress ($U_y|_{\mathrm{wall}}$) in the controlled flow can be also observed in Figure 4.18. Our boundary feedback control (tangential actuation) adjusts the flow field near the upper boundary such that the controlled wall shear stress almost matches that of the steady state profile. The region is at the edge of a small recirculation bubble (Figure 4.17) of the uncontrolled flow, hence there are some flow vectors pointing in the upstream direction while others are oriented downstream. The time is relatively short ($t = 120$) after the introduction of the control and the region is small. As a result it is still possible to see actuation both downstream and upstream. Nevertheless the controlled velocity varies continuously. Figure 4.17 shows that the effect of control is to smear the vortical structures out in the streamwise direction. It is well known that in wall bounded turbulence instabilities are generated at the wall. In two dimensional flows these instabilities are also confined to the walls. As a result, our control effectively stabilizes the flow.

We obtain approximately 71% drag reduction (see Figure 4.19) as a byproduct of our special control law. The drag in the controlled case "undershoots" below the level corresponding to the laminar flow and eventually agrees with it up to two decimal places. It is striking that even though drag reduction was not an explicit control objective (as in most of the works in this field), the stabilization

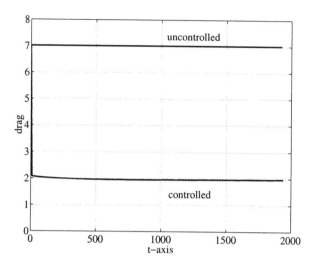

Figure 4.19: Instantaneous drag [16].

objective results in a controller that reacts to the wall shear stress error, and leads to an almost instantaneous reduction of drag to the laminar level.

4.3.3 3D Channel Flow

It is recognized that channel flow instability mechanisms are inherently 3D. Efforts that study the stabilization problem only in 2D are thus inconclusive about physical flows, for which 3D effects are quite significant. However, the "model problem" of 2D channel flow stabilization is a useful testbed for techniques that can eventually be extended to 3D flows. The Lyapunov stability analysis presented in the previous section can be extended to the 3D channel flow in a straight forward manner, and similar boundary control laws can be derived. Balogh [18] reports that, in numerical simulations, the very simple, and fully decentralized boundary control law

$$U(x,-1,z,t) = k\left(\frac{\partial U}{\partial y}(x,-1,z,t) - \frac{\partial \tilde{U}}{\partial y}(-1)\right)$$

$$U(x,1,z,t) = -k\left(\frac{\partial U}{\partial y}(x,1,z,t) - \frac{\partial \tilde{U}}{\partial y}(1)\right)$$

$$V(x,-1,z,t) = V(x,1,z,t) = 0$$

$$W(x,-1,z,t) = W(x,1,z,t) = 0$$

with $k > 0$, relaminarizes a turbulent 3D channel flow at $Re = 4000$. Figure 4.20 shows the perturbation energy as a function of time. The perturbation

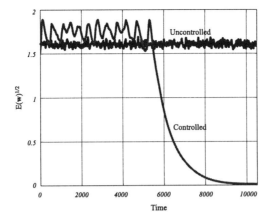

Figure 4.20: Perturbation energy (square root of) as a function of time for uncontrolled and controlled 3D channel flow at $Re = 4000$ [18].

energy is in this case

$$E(\mathbf{w}) = \int\limits_{0}^{L_z} \int\limits_{-1}^{1} \int\limits_{0}^{L_x} \left(u^2 + v^2 + w^2 \right) dx\,dy\,dz$$

and the channel dimensions are $L_x = 6\pi$ and $L_z = 3\pi$. The plots in Figure 4.21 show places where the discriminant[3] has values larger than 0.9 at $t = 4000$, indicating locations of vortical structures, and the plots in Figure 4.22 show the spanwise vorticity at $t = 5000$. It is interesting to notice that the perturbation energy is still large at these points in time (see Figure 4.20), but the control has clearly influenced the vortical structures in the flow. Figure 4.23 shows drag as a function of time. Drag is reduced to below laminar level (which is 4) almost instantly, and then gradually approaches the laminar level. This is the same result as obtained in the 2D case. Notice that drag reduction is not explicitly the objective of the control, stabilization of the parabolic equilibrium profile is. Reduction of drag to laminar level is therefore expected in the limit as $t \to \infty$, but the striking result of immediate drag reduction is obtained!

---

[3] The discriminant of the velocity gradient tensor is a scalar quantity that is commonly used in visualizations to pinpoint vortex-type motions in the flow. The following definition of the discrimant is taken from [27]: $D = (27/4)\,R^2 - Q^3$, where Q and R are the second and third invariants of the velocity gradient tensor A, defined by $Q = \frac{1}{2} \left\{ trace\,(A)^2 - trace\,(A^2) \right\}$, and $R = -det\,(A)$, respectively.

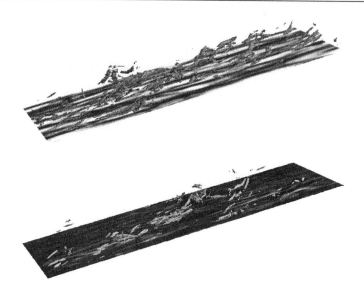

Figure 4.21: Places where the discriminant has values larger than 0.9 at $t = 4000$, for uncontrolled (upper) and controlled (lower) 3D channel flow at $Re = 4000$ [18].

4.3.4   3D Pipe Flow

In this section we show that the Lyapunov analysis that was developed for 2D and 3D channel flow, can be modified to apply for the 3D pipe flow as well.

Lyapunov Stability Analysis

Consider the following Lyapunov function candidate

$$E(\mathbf{w}) \triangleq \lim_{\varepsilon \to 0} E_\varepsilon(\mathbf{w}) \triangleq \lim_{\varepsilon \to 0} \frac{1}{2} \int_0^L \int_0^{2\pi} \int_\varepsilon^1 \left(v_r^2 + v_\theta^2 + v_z^2 \right) r\,dr\,d\theta\,dz$$

which is simply the perturbation energy of the flow. Its time derivative along trajectories of (2.65)–(2.67) is

$$\dot{E}_\varepsilon(\mathbf{w}) = \int_0^L \int_0^{2\pi} \int_\varepsilon^1 \left(v_r \dot{v}_r + v_\theta \dot{v}_\theta + v_z \dot{v}_z \right) r\,dr\,d\theta\,dz =$$

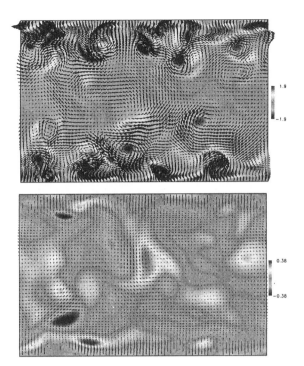

Figure 4.22: Spanwise vorticity at $t = 5000$ for uncontrolled (upper) and controlled (lower) 3D channel flow at $Re = 4000$ [18].

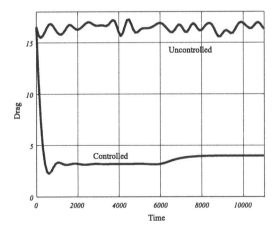

Figure 4.23: Drag as a function of time for uncontrolled and controlled 3D channel flow at $Re = 4000$ [18].

$$\int_0^L \int_0^{2\pi} \int_\varepsilon^1 \left[-\left(v_r^2 r \frac{\partial v_r}{\partial r} + v_r v_\theta \frac{\partial v_r}{\partial \theta} - v_r v_\theta^2 + r v_r \left(v_z + \tilde{V}_z \right) \frac{\partial v_r}{\partial z} \right) - r v_r \frac{\partial p}{\partial r} \right.$$

$$\left. + \frac{1}{Re} \left(r v_r \frac{\partial}{\partial r} \left(\frac{1}{r} \frac{\partial}{\partial r} (r v_r) \right) + \frac{v_r}{r} \frac{\partial^2 v_r}{\partial \theta^2} - \frac{2 v_r}{r} \frac{\partial v_\theta}{\partial \theta} + r v_r \frac{\partial^2 v_r}{\partial z^2} \right) \right] dr d\theta dz$$

$$+ \int_0^L \int_0^{2\pi} \int_\varepsilon^1 \left[-\left(r v_\theta v_r \frac{\partial v_\theta}{\partial r} + v_\theta^2 \frac{\partial v_\theta}{\partial \theta} + v_r v_\theta^2 + r v_\theta \left(v_z + \tilde{V}_z \right) \frac{\partial v_\theta}{\partial z} \right) - v_\theta \frac{\partial p}{\partial \theta} \right.$$

$$\left. + \frac{1}{Re} \left(r v_\theta \frac{\partial}{\partial r} \left(\frac{1}{r} \frac{\partial}{\partial r} (r v_\theta) \right) + \frac{v_\theta}{r} \frac{\partial^2 v_\theta}{\partial \theta^2} + \frac{2 v_\theta}{r} \frac{\partial v_r}{\partial \theta} + r v_\theta \frac{\partial^2 v_\theta}{\partial z^2} \right) \right] dr d\theta dz$$

$$+ \int_0^L \int_0^{2\pi} \int_\varepsilon^1 \left[-\left(r v_z v_r \frac{\partial \left(v_z + \bar{V}_z \right)}{\partial r} + v_z v_\theta \frac{\partial v_z}{\partial \theta} + r v_z \left(v_z + \tilde{V}_z \right) \frac{\partial v_z}{\partial z} \right) \right.$$

$$\left. - r v_z \frac{\partial p}{\partial z} + \frac{1}{Re} \left(v_z \frac{\partial}{\partial r} \left(r \frac{\partial v_z}{\partial r} \right) + v_z \frac{1}{r} \frac{\partial^2 v_z}{\partial \theta^2} + r v_z \frac{\partial^2 v_z}{\partial z^2} \right) \right] dr d\theta dz. \quad (4.218)$$

Integration by parts of the first integral in (4.218) yields

$$\int_0^L \int_0^{2\pi} \int_\varepsilon^1 \left[-\left(r v_r^2 \frac{\partial v_r}{\partial r} + v_r v_\theta \frac{\partial v_r}{\partial \theta} - v_r v_\theta^2 + r v_r \left(v_z + \tilde{V}_z \right) \frac{\partial v_r}{\partial z} \right) - r v_r \frac{\partial p}{\partial r} \right.$$

$$\left. + \frac{1}{Re} \left(r v_r \frac{\partial}{\partial r} \left(\frac{1}{r} \frac{\partial}{\partial r} (r v_r) \right) + \frac{v_r}{r} \frac{\partial^2 v_r}{\partial \theta^2} - \frac{2 v_r}{r} \frac{\partial v_\theta}{\partial \theta} + r v_r \frac{\partial^2 v_r}{\partial z^2} \right) \right] dr d\theta dz$$

$$= \int_0^L \int_0^{2\pi} \left[-\frac{1}{2} r v_r^3 \right]_\varepsilon^1 d\theta dz + \frac{1}{2} \int_0^L \int_0^{2\pi} \int_\varepsilon^1 v_r^2 \frac{\partial (r v_r)}{\partial r} dr d\theta dz$$

$$+ \frac{1}{2} \int_0^L \int_0^{2\pi} \int_\varepsilon^1 v_r^2 \frac{\partial v_\theta}{\partial \theta} dr d\theta dz + \int_0^L \int_0^{2\pi} \int_\varepsilon^1 v_r v_\theta^2 dr d\theta dz$$

$$+ \frac{1}{2} \int_0^L \int_0^{2\pi} \int_\varepsilon^1 r v_r^2 \frac{\partial v_z}{\partial z} dr d\theta dz + \int_0^L \int_0^{2\pi} \left[-r v_r p \right]_\varepsilon^1 d\theta dz$$

$$+ \int_0^L \int_0^{2\pi} \int_\varepsilon^1 \frac{\partial (r v_r)}{\partial r} p \, dr d\theta dz + \frac{1}{Re} \int_0^L \int_0^{2\pi} \left[v_r \frac{\partial}{\partial r} (r v_r) \right]_\varepsilon^1 d\theta dz$$

$$- \frac{1}{Re} \int_0^L \int_0^{2\pi} \int_\varepsilon^1 \frac{1}{r} \left(\frac{\partial}{\partial r} (r v_r) \right)^2 dr d\theta dz - \frac{1}{Re} \int_0^L \int_0^{2\pi} \int_\varepsilon^1 \frac{1}{r} \left(\frac{\partial v_r}{\partial \theta} \right)^2 dr d\theta dz$$

$$- \frac{1}{Re} \int_0^L \int_0^{2\pi} \int_\varepsilon^1 \frac{2}{r} v_r \frac{\partial v_\theta}{\partial \theta} dr d\theta dz - \frac{1}{Re} \int_0^L \int_0^{2\pi} \int_\varepsilon^1 r \left(\frac{\partial v_r}{\partial z} \right)^2 dr d\theta dz.$$

Integration by parts of the second integral in (4.218) yields

$$
\int_0^L \int_0^{2\pi} \int_\varepsilon^1 \left[-\left(r v_\theta v_r \frac{\partial v_\theta}{\partial r} + v_\theta^2 \frac{\partial v_\theta}{\partial \theta} + v_r v_\theta^2 + r v_\theta \left(v_z + \tilde{V}_z \right) \frac{\partial v_\theta}{\partial z} \right) - v_\theta \frac{\partial p}{\partial \theta} \right.
$$
$$
\left. + \frac{1}{Re} \left(r v_\theta \frac{\partial}{\partial r} \left(\frac{1}{r} \frac{\partial}{\partial r} (r v_\theta) \right) + \frac{v_\theta}{r} \frac{\partial^2 v_\theta}{\partial \theta^2} + \frac{2 v_\theta}{r} \frac{\partial v_r}{\partial \theta} + r v_\theta \frac{\partial^2 v_\theta}{\partial z^2} \right) \right] dr d\theta dz
$$
$$
= \frac{1}{2} \int_0^L \int_0^{2\pi} \left[-r v_r v_\theta^2 \right]_\varepsilon^1 d\theta dz + \frac{1}{2} \int_0^L \int_0^{2\pi} \int_\varepsilon^1 v_\theta^2 \frac{\partial (r v_r)}{\partial r} dr d\theta dz
$$
$$
+ \frac{1}{2} \int_0^L \int_0^{2\pi} \int_\varepsilon^1 v_\theta^2 \frac{\partial v_\theta}{\partial \theta} dr d\theta dz - \int_0^L \int_0^{2\pi} \int_\varepsilon^1 v_r v_\theta^2 dr d\theta dz
$$
$$
+ \frac{1}{2} \int_0^L \int_0^{2\pi} \int_\varepsilon^1 r v_\theta^2 \frac{\partial v_z}{\partial z} dr d\theta dz + \int_0^L \int_0^{2\pi} \int_\varepsilon^1 \frac{\partial v_\theta}{\partial \theta} p \, dr d\theta dz
$$
$$
+ \frac{1}{Re} \int_0^L \int_0^{2\pi} \left[v_\theta \frac{\partial}{\partial r} (r v_\theta) \right]_\varepsilon^1 d\theta dz - \frac{1}{Re} \int_0^L \int_0^{2\pi} \int_\varepsilon^1 \frac{1}{r} \left(\frac{\partial}{\partial r} (r v_\theta) \right)^2 dr d\theta dz
$$
$$
- \frac{1}{Re} \int_0^L \int_0^{2\pi} \int_\varepsilon^1 \frac{1}{r} \left(\frac{\partial v_\theta}{\partial \theta} \right)^2 dr d\theta dz + \frac{1}{Re} \int_0^L \int_0^{2\pi} \int_\varepsilon^1 \frac{2}{r} v_\theta \frac{\partial v_r}{\partial \theta} dr d\theta dz
$$
$$
- \frac{1}{Re} \int_0^L \int_0^{2\pi} \int_\varepsilon^1 r \left(\frac{\partial v_\theta}{\partial z} \right)^2 dr d\theta dz.
$$

Integration by parts of the third integral in (4.218) yields

$$
\int_0^L \int_0^{2\pi} \int_\varepsilon^1 \left[-\left(r v_z v_r \frac{\partial \left(v_z + \tilde{V}_z \right)}{\partial r} + v_z v_\theta \frac{\partial v_z}{\partial \theta} + r v_z \left(v_z + \tilde{V}_z \right) \frac{\partial v_z}{\partial z} \right) \right.
$$
$$
\left. - r v_z \frac{\partial p}{\partial z} + \frac{1}{Re} \left(v_z \frac{\partial}{\partial r} \left(r \frac{\partial v_z}{\partial r} \right) + v_z \frac{1}{r} \frac{\partial^2 v_z}{\partial \theta^2} + r v_z \frac{\partial^2 v_z}{\partial z^2} \right) \right] dr d\theta dz
$$
$$
= \frac{1}{2} \int_0^L \int_0^{2\pi} \left[-r v_z^2 v_r \right]_\varepsilon^1 d\theta dz + \frac{1}{2} \int_0^L \int_0^{2\pi} \int_\varepsilon^1 v_z^2 \frac{\partial (r v_r)}{\partial r} dr d\theta dz
$$
$$
- \int_0^L \int_0^{2\pi} \int_\varepsilon^1 r v_z v_r \frac{\partial \tilde{V}_z}{\partial r} dr d\theta dz + \frac{1}{2} \int_0^L \int_0^{2\pi} \int_\varepsilon^1 v_z^2 \frac{\partial v_\theta}{\partial \theta} dr d\theta dz
$$
$$
+ \frac{1}{2} \int_0^L \int_0^{2\pi} \int_\varepsilon^1 r v_z^2 \frac{\partial v_z}{\partial z} dr d\theta dz + \int_0^L \int_0^{2\pi} \int_\varepsilon^1 r \frac{\partial v_z}{\partial z} p \, dr d\theta dz
$$

$$+ \frac{1}{Re} \int_0^L \int_0^{2\pi} \left[r v_z \frac{\partial v_z}{\partial r} \right]_\varepsilon^1 d\theta dz - \frac{1}{Re} \int_0^L \int_0^{2\pi} \int_\varepsilon^1 r \left(\frac{\partial v_z}{\partial r} \right)^2 dr d\theta dz$$

$$- \frac{1}{Re} \int_0^L \int_0^{2\pi} \int_\varepsilon^1 \frac{1}{r} \left(\frac{\partial v_z}{\partial \theta} \right)^2 dr d\theta dz - \frac{1}{Re} \int_0^L \int_0^{2\pi} \int_\varepsilon^1 r \left(\frac{\partial v_z}{\partial z} \right)^2 dr d\theta dz.$$

Inserting the results into (4.218) yields

$$\dot{E}_\varepsilon \left(\mathbf{w} \right) =$$

$$\frac{1}{2} \int_0^L \int_0^{2\pi} \int_\varepsilon^1 \left(v_r^2 + v_\theta^2 + v_z^2 + \frac{2}{\rho} p \right) \underbrace{\left(\frac{1}{r} \frac{\partial \left(r v_r \right)}{\partial r} + \frac{1}{r} \frac{\partial v_\theta}{\partial \theta} + \frac{\partial v_z}{\partial z} \right)}_{=0 \text{ by incompressibility}} r dr d\theta dz$$

$$+ \int_0^L \int_0^{2\pi} \left[-\frac{1}{2} r v_r^3 \right]_\varepsilon^1 d\theta dz + \int_0^L \int_0^{2\pi} \left[-r v_r p \right]_\varepsilon^1 d\theta dz + \frac{1}{Re} \int_0^L \int_0^{2\pi} \left[v_r \frac{\partial}{\partial r} \left(r v_r \right) \right]_\varepsilon^1 d\theta dz$$

$$+ \frac{1}{2} \int_0^L \int_0^{2\pi} \left[-r v_r v_\theta^2 \right]_\varepsilon^1 d\theta dz + \frac{1}{Re} \int_0^L \int_0^{2\pi} \left[v_\theta \frac{\partial}{\partial r} \left(r v_\theta \right) \right]_\varepsilon^1 d\theta dz$$

$$+ \frac{1}{2} \int_0^L \int_0^{2\pi} \left[-r v_z^2 v_r \right]_\varepsilon^1 d\theta dz + \frac{1}{Re} \int_0^L \int_0^{2\pi} \left[r v_z \frac{\partial v_z}{\partial r} \right]_\varepsilon^1 d\theta dz$$

$$- \int_0^L \int_0^{2\pi} \int_\varepsilon^1 r v_z v_r \frac{\partial \tilde{V}_z}{\partial r} dr d\theta dz - \frac{1}{Re} \int_0^L \int_0^{2\pi} \int_\varepsilon^1 \frac{2}{r} v_r \frac{\partial v_\theta}{\partial \theta} dr d\theta dz$$

$$+ \frac{1}{Re} \int_0^L \int_0^{2\pi} \int_\varepsilon^1 \frac{2}{r} v_\theta \frac{\partial v_r}{\partial \theta} dr d\theta dz - \frac{1}{Re} \int_0^L \int_0^{2\pi} \int_\varepsilon^1 \frac{1}{r} \left(\frac{\partial}{\partial r} \left(r v_r \right) \right)^2 dr d\theta dz$$

$$- \frac{1}{Re} \int_0^L \int_0^{2\pi} \int_\varepsilon^1 \frac{1}{r} \left(\frac{\partial v_r}{\partial \theta} \right)^2 dr d\theta dz - \frac{1}{Re} \int_0^L \int_0^{2\pi} \int_\varepsilon^1 r \left(\frac{\partial v_r}{\partial z} \right)^2 dr d\theta dz$$

$$- \frac{1}{Re} \int_0^L \int_0^{2\pi} \int_\varepsilon^1 \frac{1}{r} \left(\frac{\partial}{\partial r} \left(r v_\theta \right) \right)^2 dr d\theta dz - \frac{1}{Re} \int_0^L \int_0^{2\pi} \int_\varepsilon^1 \frac{1}{r} \left(\frac{\partial v_\theta}{\partial \theta} \right)^2 dr d\theta dz$$

$$- \frac{1}{Re} \int_0^L \int_0^{2\pi} \int_\varepsilon^1 r \left(\frac{\partial v_\theta}{\partial z} \right)^2 dr d\theta dz - \frac{1}{Re} \int_0^L \int_0^{2\pi} \int_\varepsilon^1 r \left(\frac{\partial v_z}{\partial r} \right)^2 dr d\theta dz$$

$$- \frac{1}{Re} \int_0^L \int_0^{2\pi} \int_\varepsilon^1 \frac{1}{r} \left(\frac{\partial v_z}{\partial \theta} \right)^2 dr d\theta dz - \frac{1}{Re} \int_0^L \int_0^{2\pi} \int_\varepsilon^1 r \left(\frac{\partial v_z}{\partial z} \right)^2 dr d\theta dz.$$

Since

$$\int_0^L \int_0^{2\pi} \int_\varepsilon^1 v_r \frac{\partial v_r}{\partial r} dr d\theta dz = \frac{1}{2} \int_0^L \int_0^{2\pi} \left[v_r^2 \right]_{r=\varepsilon}^1 d\theta dz$$

$$\int_0^L \int_0^{2\pi} \int_\varepsilon^1 v_\theta \frac{\partial v_\theta}{\partial r} dr d\theta dz = \frac{1}{2} \int_0^L \int_0^{2\pi} \left[v_\theta^2 \right]_{r=\varepsilon}^1 d\theta dz$$

we have that

$$-\frac{1}{Re} \int_0^L \int_0^{2\pi} \int_\varepsilon^1 \frac{1}{r} \left(\frac{\partial}{\partial r} (r v_r) \right)^2 dr d\theta dz - \frac{1}{Re} \int_0^L \int_0^{2\pi} \int_\varepsilon^1 \frac{1}{r} \left(\frac{\partial}{\partial r} (r v_\theta) \right)^2 dr d\theta dz$$

$$-\frac{1}{Re} \int_0^L \int_0^{2\pi} \int_\varepsilon^1 \frac{1}{r} \left(\frac{\partial v_r}{\partial \theta} \right)^2 dr d\theta dz - \frac{1}{Re} \int_0^L \int_0^{2\pi} \int_\varepsilon^1 \frac{1}{r} \left(\frac{\partial v_\theta}{\partial \theta} \right)^2 dr d\theta dz$$

$$+\frac{1}{Re} \int_0^L \int_0^{2\pi} \int_\varepsilon^1 \frac{2}{r} v_\theta \frac{\partial v_r}{\partial \theta} dr d\theta dz - \frac{1}{Re} \int_0^L \int_0^{2\pi} \int_\varepsilon^1 \frac{2}{r} v_r \frac{\partial v_\theta}{\partial \theta} dr d\theta dz$$

$$= -\frac{1}{Re} \int_0^L \int_0^{2\pi} \int_\varepsilon^1 \frac{1}{r} \left(v_r^2 + 2 r v_r \frac{\partial v_r}{\partial r} + r^2 \left(\frac{\partial v_r}{\partial r} \right)^2 \right) dr d\theta dz$$

$$-\frac{1}{Re} \int_0^L \int_0^{2\pi} \int_\varepsilon^1 \frac{1}{r} \left(v_\theta^2 + 2 r v_\theta \frac{\partial v_\theta}{\partial r} + r^2 \left(\frac{\partial v_\theta}{\partial r} \right)^2 \right) dr d\theta dz$$

$$-\frac{1}{Re} \int_0^L \int_0^{2\pi} \int_\varepsilon^1 \frac{1}{r} \left(\frac{\partial v_r}{\partial \theta} \right)^2 dr d\theta dz - \frac{1}{Re} \int_0^L \int_0^{2\pi} \int_\varepsilon^1 \frac{1}{r} \left(\frac{\partial v_\theta}{\partial \theta} \right)^2 dr d\theta dz$$

$$+\frac{1}{Re} \int_0^L \int_0^{2\pi} \int_\varepsilon^1 \frac{2}{r} v_\theta \frac{\partial v_r}{\partial \theta} dr d\theta dz - \frac{1}{Re} \int_0^L \int_0^{2\pi} \int_\varepsilon^1 \frac{2}{r} v_r \frac{\partial v_\theta}{\partial \theta} dr d\theta dz$$

$$= -\frac{1}{Re} \int_0^L \int_0^{2\pi} \left[v_r^2 + v_\theta^2 \right]_{r=\varepsilon}^1 d\theta dz - \frac{1}{Re} \int_0^L \int_0^{2\pi} \int_\varepsilon^1 \frac{1}{r} \left(v_r^2 + r^2 \left(\frac{\partial v_r}{\partial r} \right)^2 \right) dr d\theta dz$$

$$-\frac{1}{Re} \int_0^L \int_0^{2\pi} \int_\varepsilon^1 \frac{1}{r} \left(v_\theta^2 + r^2 \left(\frac{\partial v_\theta}{\partial r} \right)^2 \right) dr d\theta dz - \frac{1}{Re} \int_0^L \int_0^{2\pi} \int_\varepsilon^1 \frac{1}{r} \left(\frac{\partial v_r}{\partial \theta} \right)^2 dr d\theta dz$$

$$-\frac{1}{Re} \int_0^L \int_0^{2\pi} \int_\varepsilon^1 \frac{1}{r} \left(\frac{\partial v_\theta}{\partial \theta} \right)^2 dr d\theta dz + \frac{1}{Re} \int_0^L \int_0^{2\pi} \int_\varepsilon^1 \frac{2}{r} v_\theta \frac{\partial v_r}{\partial \theta} dr d\theta dz$$

$$-\frac{1}{Re}\int_0^L\int_0^{2\pi}\int_\varepsilon^1\frac{2}{r}v_r\frac{\partial v_\theta}{\partial\theta}drd\theta dz$$

$$=-\frac{1}{Re}\int_0^L\int_0^{2\pi}\left[v_r^2+v_\theta^2\right]_{r=\varepsilon}^1 d\theta dz-\frac{1}{Re}\int_0^L\int_0^{2\pi}\int_\varepsilon^1 r\left(\frac{\partial v_r}{\partial r}\right)^2 drd\theta dz$$

$$-\frac{1}{Re}\int_0^L\int_0^{2\pi}\int_\varepsilon^1 r\left(\frac{\partial v_\theta}{\partial r}\right)^2 drd\theta dz-\frac{1}{Re}\int_0^L\int_0^{2\pi}\int_\varepsilon^1\frac{1}{r}\left(v_r+\frac{\partial v_\theta}{\partial\theta}\right)^2 drd\theta dz$$

$$-\frac{1}{Re}\int_0^L\int_0^{2\pi}\int_\varepsilon^1\frac{1}{r}\left(v_\theta-\frac{\partial v_r}{\partial\theta}\right)^2 drd\theta dz.$$

Therefore, we get

$$\dot{E}_\varepsilon(\mathbf{w})=\int_0^L\int_0^{2\pi}\left[-\frac{1}{2}rv_r^3\right]_\varepsilon^1 d\theta dz+\int_0^L\int_0^{2\pi}\left[-rv_r p\right]_\varepsilon^1 d\theta dz$$

$$+\frac{1}{Re}\int_0^L\int_0^{2\pi}\left[v_r\frac{\partial}{\partial r}(rv_r)\right]_\varepsilon^1 d\theta dz+\frac{1}{2}\int_0^L\int_0^{2\pi}\left[-rv_r v_\theta^2\right]_\varepsilon^1 d\theta dz$$

$$+\frac{1}{Re}\int_0^L\int_0^{2\pi}\left[v_\theta\frac{\partial}{\partial r}(rv_\theta)\right]_\varepsilon^1 d\theta dz+\frac{1}{2}\int_0^L\int_0^{2\pi}\left[-rv_z^2 v_r\right]_\varepsilon^1 d\theta dz$$

$$+\frac{1}{Re}\int_0^L\int_0^{2\pi}\left[rv_z\frac{\partial v_z}{\partial r}\right]_\varepsilon^1 d\theta dz-\frac{1}{Re}\int_0^L\int_0^{2\pi}\left[v_r^2+v_\theta^2\right]_{r=\varepsilon}^1 d\theta dz$$

$$-\int_0^L\int_0^{2\pi}\int_\varepsilon^1 rv_z v_r\frac{\partial\bar{V}_z}{\partial r}drd\theta dz-\frac{1}{Re}\int_0^L\int_0^{2\pi}\int_\varepsilon^1 r\left(\frac{\partial v_r}{\partial z}\right)^2 drd\theta dz$$

$$-\frac{1}{Re}\int_0^L\int_0^{2\pi}\int_\varepsilon^1 r\left(\frac{\partial v_\theta}{\partial z}\right)^2 drd\theta dz-\frac{1}{Re}\int_0^L\int_0^{2\pi}\int_\varepsilon^1 r\left(\frac{\partial v_z}{\partial r}\right)^2 drd\theta dz$$

$$-\frac{1}{Re}\int_0^L\int_0^{2\pi}\int_\varepsilon^1\frac{1}{r}\left(\frac{\partial v_z}{\partial\theta}\right)^2 drd\theta dz-\frac{1}{Re}\int_0^L\int_0^{2\pi}\int_\varepsilon^1 r\left(\frac{\partial v_z}{\partial z}\right)^2 drd\theta dz$$

$$-\frac{1}{Re}\int_0^L\int_0^{2\pi}\int_\varepsilon^1 r\left(\frac{\partial v_r}{\partial r}\right)^2 drd\theta dz-\frac{1}{Re}\int_0^L\int_0^{2\pi}\int_\varepsilon^1 r\left(\frac{\partial v_\theta}{\partial r}\right)^2 drd\theta dz$$

$$-\frac{1}{Re}\int\limits_0^L\int\limits_0^{2\pi}\int\limits_\varepsilon^1 \frac{1}{r}\left(v_r+\frac{\partial v_\theta}{\partial \theta}\right)^2 drd\theta dz -\frac{1}{Re}\int\limits_0^L\int\limits_0^{2\pi}\int\limits_\varepsilon^1 \frac{1}{r}\left(v_\theta-\frac{\partial v_r}{\partial \theta}\right)^2 drd\theta dz.$$

$$(4.219)$$

Now, we write

$$v_r(r,\theta,z,t) = v_r(1,\theta,z,t) - \int\limits_r^1 \frac{\partial v_r}{\partial r} dr$$

so that

$$v_r^2(r,\theta,z,t) = \left(v_r(1,\theta,z,t) - \int\limits_r^1 \frac{\partial v_r}{\partial r} dr\right)^2 \leq 2v_r^2(1,\theta,z,t)+2\left(\int\limits_r^1 \frac{\partial v_r}{\partial r} dr\right)^2.$$

By the Schwartz inequality,

$$\left(\int\limits_r^1 \frac{1}{\sqrt{r}}\sqrt{r}\frac{\partial v_r}{\partial r} dr\right)^2 \leq -\ln r \int\limits_r^1 r\left(\frac{\partial v_r}{\partial r}\right)^2 dr$$

so we have that

$$rv_r^2(r,\theta,z,t) \leq 2rv_r^2(1,\theta,z,t) - 2r\ln r \int\limits_\varepsilon^1 r\left(\frac{\partial v_r}{\partial r}\right)^2 dr$$

where we have set $r=\varepsilon$ in the lower integral limit. We now get

$$\int\limits_0^L\int\limits_0^{2\pi}\int\limits_\varepsilon^1 v_r^2 r drd\theta dz \leq$$

$$\int\limits_0^L\int\limits_0^{2\pi}\int\limits_\varepsilon^1 \left(2rv_r^2(1,\theta,z,t) - 2r\ln r \int\limits_\varepsilon^1 r\left(\frac{\partial v_r}{\partial r}\right)^2 dr\right) drd\theta dz$$

$$=\left(1-\varepsilon^2\right)\int\limits_0^L\int\limits_0^{2\pi} v_r^2(1,\theta,z,t) d\theta dz$$

$$+\frac{1}{2}\left(1-\varepsilon^2+2\varepsilon^2\ln\varepsilon\right)\int\limits_0^L\int\limits_0^{2\pi}\int\limits_\varepsilon^1 r\left(\frac{\partial v_r}{\partial r}\right)^2 drd\theta dz.$$

Analogous derivations for v_θ, and v_z, yield

$$-\frac{1}{Re}\int\limits_0^L\int\limits_0^{2\pi}\int\limits_\varepsilon^1\left(\left(\frac{\partial v_r}{\partial r}\right)^2+\left(\frac{\partial v_\theta}{\partial r}\right)^2+\left(\frac{\partial v_z}{\partial r}\right)^2\right)r\,dr\,d\theta\,dz\leq$$

$$-\frac{4}{Re\left(1-\varepsilon^2+2\varepsilon^2\ln\varepsilon\right)}E(\mathbf{w})$$

$$+\frac{2\left(1-\varepsilon^2\right)}{Re\left(1-\varepsilon^2+2\varepsilon^2\ln\varepsilon\right)}\int\limits_0^L\int\limits_0^{2\pi}\left(v_r^2\left(1,\theta,z,t\right)+v_\theta^2\left(1,\theta,z,t\right)+v_z^2\left(1,\theta,z,t\right)\right)d\theta\,dz.$$

Inserting this result into (4.219), and letting $\varepsilon\to0$, gives

$$\dot{E}(\mathbf{w})\leq-2\left(\frac{2}{Re}-1\right)E(\mathbf{w})+\frac{2}{Re}\int\limits_0^L\int\limits_0^{2\pi}\left[v_r^2+v_\theta^2+v_z^2\right]_{r=1}d\theta\,dz$$

$$-\frac{1}{2}\int\limits_0^L\int\limits_0^{2\pi}\left[v_r^3\right]_{r=1}d\theta\,dz-\int\limits_0^L\int\limits_0^{2\pi}\left[v_r p\right]_{r=1}d\theta\,dz-\frac{1}{2}\int\limits_0^L\int\limits_0^{2\pi}\left[v_r v_\theta^2\right]_{r=1}d\theta\,dz$$

$$-\frac{1}{2}\int\limits_0^L\int\limits_0^{2\pi}\left[v_z^2 v_r\right]_{r=1}d\theta\,dz+\frac{1}{Re}\int\limits_0^L\int\limits_0^{2\pi}\left[v_r\frac{\partial v_r}{\partial r}\right]_{r=1}d\theta\,dz$$

$$+\frac{1}{Re}\int\limits_0^L\int\limits_0^{2\pi}\left[v_\theta\frac{\partial v_\theta}{\partial r}\right]_{r=1}d\theta\,dz+\frac{1}{Re}\int\limits_0^L\int\limits_0^{2\pi}\left[v_z\frac{\partial v_z}{\partial r}\right]_{r=1}d\theta\,dz\quad(4.220)$$

where we have used

$$-\int\limits_0^L\int\limits_0^{2\pi}\int\limits_\varepsilon^1 r v_z v_r\frac{\partial\tilde{V}_z}{\partial r}dr\,d\theta\,dz\leq\int\limits_0^L\int\limits_0^{2\pi}\int\limits_\varepsilon^1 r\left|v_z\right|\left|v_r\right|\left|\frac{\partial\tilde{V}_z}{\partial r}\right|dr\,d\theta\,dz$$

$$\leq\int\limits_0^L\int\limits_0^{2\pi}\int\limits_0^1\left(v_r^2+v_\theta^2+v_z^2\right)r\,dr\,d\theta\,dz=2E(\mathbf{w})$$

and

$$\frac{1}{Re}\int\limits_0^L\int\limits_0^{2\pi}\left[v_r\frac{\partial}{\partial r}\left(r v_r\right)\right]_\varepsilon^1 d\theta\,dz+\frac{1}{Re}\int\limits_0^L\int\limits_0^{2\pi}\left[v_\theta\frac{\partial}{\partial r}\left(r v_\theta\right)\right]_\varepsilon^1 d\theta\,dz$$

$$-\frac{1}{Re}\int\limits_0^L\int\limits_0^{2\pi}\left[v_r^2+v_\theta^2\right]_\varepsilon^1 d\theta\,dz$$

$$= \frac{1}{Re} \int_0^L \int_0^{2\pi} \left[v_r \frac{\partial v_r}{\partial r} \right]_{r=1} d\theta dz + \frac{1}{Re} \int_0^L \int_0^{2\pi} \left[v_\theta \frac{\partial v_\theta}{\partial r} \right]_{r=1} d\theta dz.$$

In (4.220), all but the first term on the right hand side of the inequality are evaluated on the boundary. These are the terms by which boundary control laws are designed. Below, a pressure-based control law is presented.

Boundary Control

We propose the following control law

$$v_r(1, \theta, z, t) = k(p(1, \theta, z, t) - p(1, \theta + \pi, z, t))$$
$$v_\theta (1, \theta, z, t) = 0 \qquad\qquad (4.221)$$
$$v_z (1, \theta, z, t) = 0.$$

Inserting into (4.220), yields

$$\dot{E}(\mathbf{w}) \le -2 \left(\frac{2}{Re} - 1 \right) E(\mathbf{w}) + \frac{1}{Re} \int_0^L \int_0^{2\pi} v_r^2 (1, \theta, z, t) \, d\theta dz$$

$$- \frac{1}{2} \int_0^L \int_0^{2\pi} v_r^3 (1, \theta, z, t) \, d\theta dz - \int_0^L \int_0^{2\pi} v_r (1, \theta, z, t) p(1, \theta, z, t) \, d\theta dz \qquad (4.222)$$

For n odd, we get

$$\int_0^{2\pi} v_r^n (1, \theta, z, t) d\theta = \int_{-\pi}^0 v_r^n (1, \theta, z, t) \, d\theta + \int_0^\pi v_r^n (1, \theta, z, t) \, d\theta$$

$$= - \int_{-\pi}^0 v_r^n (1, -\theta, z, t) \, d\theta + \int_0^\pi v_r^n (1, \theta, z, t) \, d\theta$$

and by a change of variables in the first integral ($\theta^* = -\theta$), we get

$$\int_0^{2\pi} v_r^n (1, \theta, z) d\theta = - \int_0^\pi v_r^n (1, \theta^*, z, t) \, d\theta^* + \int_0^\pi v_r^n (1, \theta, z, t) \, d\theta = 0$$

which proves that the net mass flow through the wall is zero, and that

$$\int_0^{2\pi} [v_r^3]_{r=1} \, d\theta = 0$$

by setting $n = 1$ and $n = 3$, respectively. Since

$$\int_0^{2\pi} v_r\,(1,\theta,z,t)\,p\,(1,\theta,z,t)\,d\theta$$

$$= \int_{-\pi}^{0} v_r\,(1,\theta,z,t)\,p\,(1,\theta,z,t)\,d\theta + \int_0^{\pi} v_r\,(1,\theta,z,t)\,p\,(1,\theta,z,t)\,d\theta$$

$$= -\int_{\pi}^{0} v_r\,(1,-\theta^*,z,t)\,p\,(1,-\theta^*,z,t)\,d\theta^* + \int_0^{\pi} v_r\,(1,\theta,z,t)\,p\,(1,\theta,z,t)\,d\theta$$

$$= -\int_0^{\pi} v_r\,(1,\theta^*,z,t)\,p\,(1,-\theta^*,z,t)\,d\theta^* + \int_0^{\pi} v_r\,(1,\theta,z,t)\,p\,(1,\theta,z,t)\,d\theta$$

$$= \int_0^{\pi} v_r\,(1,\theta,z,t)\,(p\,(1,\theta,z,t) - p\,(1,-\theta,z,t))\,d\theta$$

$$= \frac{1}{k}\int_0^{\pi} v_r^2\,(1,\theta,z,t)\,d\theta = \frac{1}{2k}\int_0^{2\pi} v_r^2\,(1,\theta,z,t)\,d\theta$$

we obtain

$$\dot{E}(\mathbf{w}) \leq -2\left(\frac{2}{Re}-1\right)E(\mathbf{w}) - \left(\frac{1}{2k}-\frac{1}{Re}\right)\int_0^{L}\int_0^{2\pi} v_r^2\,(1,\theta,z,t)\,d\theta dz. \quad (4.223)$$

Thus, we have shown that the equilibrium profile is globally exponentially stable in L_2 for sufficiently small Re and for appropriately selected feedback gain k. Note that the control law is decentralized and has a symmetrical structure similar to that of the pressure-based control designed for the 2D channel flow case (see equations (4.63)–(4.64)).

4.3.5   Drag Reduction Below Laminar Flow

The simple pressure-based feedback control strategy for wall-transpiration control of incompressible unsteady 2D channel flow proposed in Section 4.3.1, leads to flow transients with instantaneous drag far lower than that of the corresponding laminar flow (see lower right graph in Figure 4.12). This touches at the common belief that the laminar flow constitutes a fundamental limit to the drag reduction that is possible to obtain, which is stated in [25] as the following conjecture

Conjecture 4.1 *The lowest sustainable drag of an incompressible constant mass-flux channel flow in either 2D or 3D, when controlled via a distribution of zero-net mass-flux blowing/suction over the channel walls, is exactly that of the laminar flow.*

We denote the drag of the laminar flow D_L. By sustainable drag, D_∞, we mean the time average (denoted $\bar{D}(t)$) of the instantaneous drag, $D(t)$, as the averaging time T approaches infinity, i.e.,

$$D_\infty \triangleq \lim_{T \to \infty} \bar{D}(T) \triangleq \lim_{T \to \infty} \frac{1}{T} \int_0^T D(t) dt$$

with the instantaneous drag given as in (4.72), multiplied by the factor μL. In [29] the mechanisms that initiate the $D(t) < D_L$ transient are investigated, and an attempt at sustaining the drag below that of laminar flow is made. A simulation of a constant mass-flux 2D channel flow at $Re = 7500$, using a box length 60 times the channel half width serves as an illustration. The flow at $t = 0^-$ in Figure 4.24, a fully established unsteady flow in a 2D channel (see, e.g., [77]), has extensive regions of backflow near the walls. This appears to be the key to initiating a $D(t) < D_L$ transient. By generally applying suction at the walls in regions of negative drag, and applying blowing in regions of large positive drag, the negative drag regions are intensified (locally, more negative drag) and the high positive drag regions are moderated (locally, less positive drag), as illustrated in Figure 4.25. In terms of reducing the total instantaneous drag $D(t)$ integrated over the walls at time $t = 5$, both effects are beneficial, and thus the control application results, for a brief amount of time, in a "win-win" situation, facilitating a drastic transient reduction in skin-friction drag to well below laminar levels. Unfortunately, the wall suction quickly acts to remove the backflow from the flow domain entirely, after which the instantaneous drag $D(t)$ asymptotes back to the laminar level, D_L.

A metric which quantifies the degree of backflow present at any instant in a particular flow is given by

$$b_p = \left[\frac{1}{A} \int_{\Omega^-} |U|^p d\Omega \right]^{1/p}$$

where Ω^- is the subset of the channel flow domain Ω which is characterized by regions of flow with negative streamwise velocity, i.e., $\Omega^- = \{\Omega(x, y) | U(x, y) < 0\}$, and A is the volume of the entire channel domain Ω. For the simulation depicted in Figures 4.24–4.25, plots of the history of b_1 and b_2 are shown in Figure 4.26. Note that, by both measures, the backflow is quickly eliminated after the control is initiated; flow visualizations such as Figure 4.25 demonstrate clearly that the backflowing fluid in Ω^- is simply removed from the channel by the control suction.

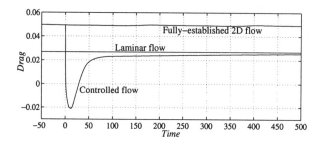

Figure 4.24: History of drag. Simulation initiated from fully established unsteady 2D flow at $Re = 7500$. Stabilizing pressure-based feedback control strategy with $k = -0.125$ turned on at $t = 0$ [29].

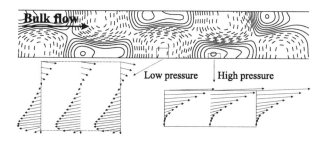

Figure 4.25: Win-win mechanism at $t = 5$: intensification of local regions of negative drag by suction in low pressure regions and moderation of positive drag by blowing in high pressure regions. Shown are contours of pressure in 1/6 of the computational domain (top) and selected velocity profiles (bottom) [29].

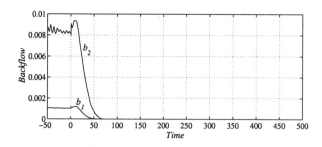

Figure 4.26: Elimination of backflow after control is turned on, as measured by $b_1(t)$ and $b_2(t)$ [29].

As a "standard" problem to test the utility of a given control strategy for reducing time-averaged drag to below laminar levels, a series of controlled 2D channel flow simulations at $Re = 7500$ were initialized from small (random) perturbations to a laminar flow profile. The control producing the $D(t) < D_L$ transients was cycled off and on periodically, with the "running average" of the drag, $\overline{D}(t) = \int_0^t D(t')\,dt'$, computed as the flow evolved to quantify progress towards sustained drag reduction. A large variety of different periods, duty cycles, and control amplitudes were explored; see [29] for details.

The results indicate that it is always necessary to pay a more expensive price (in terms of the time-averaged drag) to obtain the backflow than the benefit (in terms of the time-averaged drag) that can be obtained by applying suction to the backflow regions. Therefore, it appears that the parabolic equilibrium profile represents a fundamental limit to the drag reduction possible when applying control based on wall transpiration with zero-net mass flow.

The objective of the control strategies above has been to stabilize the parabolic equilibrium profile. Although not explicitly stated in the objective, drag reduction is obtained indirectly using this strategy. For wall normal actuation, with zero-net mass flow, the objective of stabilizing the parabolic profile is equivalent to minimizing drag, due to Conjecture 4.1. However, it is clear that drag can be reduced further if actuation is directed in the streamwise direction, and in particular if it is allowed a steady-state offset. This calls for new actuators that allow slip boundary conditions at rigid walls. Advancements in this direction based on conveyer belts as actuators, allowing slip boundary conditions, are currently being pursued [30].

4.4   Suppression of Vortex Shedding

4.4.1   Simulations of the Controlled Navier-Stokes Equation

The flow past a 2D circular cylinder has been a popular model flow for studying vortex shedding suppression by means of open-loop or feedback control. For Reynolds numbers slightly larger than the critical value for onset of vortex shedding (which is approximately $Re_c = 47$), several authors have successfully suppressed vortex shedding in numerical simulations using various simple feedback control configurations. In [113], a pair of suction/blowing slots positioned on the cylinder wall were used for actuation, and shedding was suppressed for $Re = 60$, using proportional feedback from a single velocity measurement taken some distance downstream of the cylinder. Using FLUENT[4], this result has been reproduced on a grid of approximately 4000 nodes shown in Figure 4.27.

---

[4]FLUENT is a commercial computational fluid dynamics (CFD) package available from Fluent Inc.

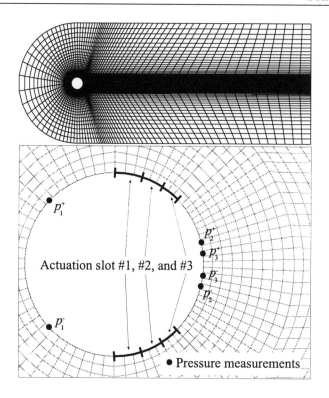

Figure 4.27: Computational grid, and zoom showing slots for blowing and suction, and locations at which pressure measurements are taken.

The initial condition for the simulations is obtained by starting from a perturbed velocity field and running the simulation for 500 time units. The result is a periodic steady state, as indicated in Figure 4.28 in terms of the lift coefficient, which is the normalized force acting on the cylinder in the vertical direction. The vorticity field at the end of this initial run is shown in Figure 4.29, where the von Kármán vortex street is clearly visible.

For the controlled flow, starting from the initial data from Figure 4.29, the time evolutions of the lift coefficient is shown in Figure 4.30, and the control effort is shown in Figure 4.31. They indicate that the vortex shedding is weakened, which is confirmed by the nearly symmetric (about the streamwise axis) vorticity map at $t = 750$ shown in Figure 4.32. The controlled velocity field continues to oscillate a little, though. In terms of the lift coefficient, it continues to vary between $\approx \pm 0.02$ beyond the time interval shown in Figure 4.30, which is about 10% of its amplitude in the uncontrolled case.

For $Re = 80$, vortex shedding was reduced, but substantially less than for the $Re = 60$ case. In [60], the same actuation configuration was tried for $Re = 60$,

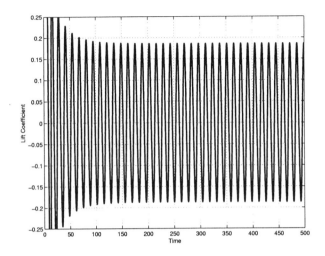

Figure 4.28: Lift coefficient for initial simulation at $Re = 60$.

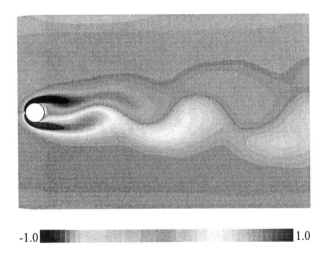

Figure 4.29: Vorticity field at $t = 500$ for $Re = 60$.

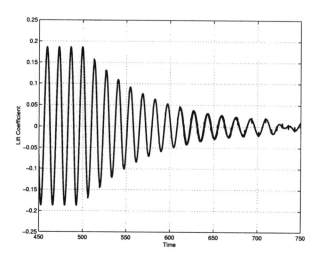

Figure 4.30: Lift coefficient for stabilizing control at $Re = 60$. Feedback from velocity measurement.

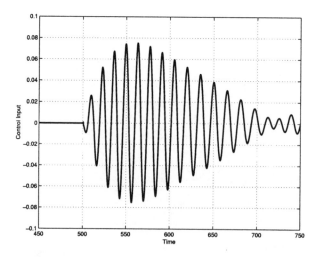

Figure 4.31: Control input for stabilizing control at $Re = 60$. Feedback from velocity measurement.

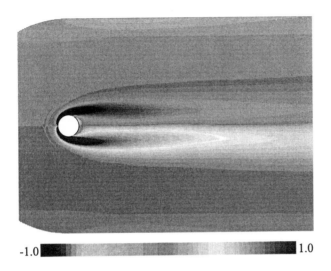

-1.0 ▮▮▮▮▮▮▮▮▮▮ 1.0

Figure 4.32: Vorticity field at $t = 750$ for stabilizing control at $Re = 60$. Feedback from velocity measurement.

using feedback from a pair of pressure sensors located on the cylinder wall. This attempt was unsuccessful, but by adding a third actuation slot, shedding was reduced considerably, even at $Re = 80$. In [6], simulations showed that stabilization at $Re = 60$ is also possible by adding more pressure sensors instead of adding an additional actuation slot.

Although some success in controlling vortex shedding has been achieved in numerical simulations, rigorous control designs are scarce due to the complexity of designing controllers based on the Navier-Stokes equation. A much simpler model, the Ginzburg-Landau equation with appropriate coefficients, has been found to model well the dynamics of vortex shedding near the critical value of the Reynolds number [71]. In [117], it was shown numerically that the Ginzburg-Landau model for Reynolds numbers close to Re_c can be stabilized using proportional feedback from a single measurement downstream of the cylinder, to local forcing at the location of the cylinder. In [95], using the model from [117], stabilization was obtained in numerical simulations for $Re = 100$, with an LQG controller designed for the linearized Ginzburg-Landau equation.

In the following sections, we present a controller designed using backstepping, that is shown to globally stabilize the equilibrium at zero of a finite difference discretization of any order of the nonlinear Ginzburg-Landau model presented in [117]. These sections are based on [4] (also in [5]), and the method is similar to the work presented in [32]. The design is valid for any Reynolds number. Numerical simulations are provided in order to demonstrate the performance of the controller, along with the potential of using low order discretizations for the control design, and thereby reducing the number of sensors needed for implementation.

4.4.2 The Ginzburg-Landau Equation

The Ginzburg-Landau equation in the notation of [117] is given by

$$\frac{\partial A}{\partial t} = -a_1(x)\frac{\partial A}{\partial x} - a_2\frac{\partial^2 A}{\partial x^2} - a_4(x)A + a_5|A|^2 A + \delta(x - x_a)u \quad (4.224)$$

where $x \in \mathbb{R}$, $A : \mathbb{R} \times \mathbb{R}_+ \to \mathbb{C}$, $a_1, a_4 : \mathbb{R} \to \mathbb{C}$, and $a_2, a_5 \in \mathbb{C}$. δ denotes the Dirac distribution and $u : \mathbb{R}_+ \to \mathbb{C}$ is the control input. Thus, control input is in the form of local forcing at x_a. The boundary conditions are $A(x \to \pm\infty, t) = 0$, that is, homogeneous Dirichlet boundary conditions. We now rewrite the equation to obtain two coupled partial differential equations in real variables and coefficients by defining

$$\rho \triangleq \frac{1}{2}(A + \bar{A}), \ \iota \triangleq \frac{1}{2i}(A - \bar{A}) \quad (4.225)$$

$$a_{R_j} \triangleq \frac{1}{2}(a_j + \bar{a}_j), \ a_{I_j} \triangleq \frac{1}{2i}(a_j - \bar{a}_j), \ j = 1, 2, 4, 5 \quad (4.226)$$

$$u_R \triangleq \frac{1}{2}(u + \bar{u}), \ u_I \triangleq \frac{1}{2i}(u - \bar{u}) \quad (4.227)$$

where i denotes the imaginary unit and $\bar{\ }$ denotes complex conjugation. With these definitions we obtain

$$\begin{aligned}
\frac{\partial \rho}{\partial t} &= \frac{1}{2}\left(\frac{\partial A}{\partial t} + \frac{\partial \bar{A}}{\partial t}\right) \\
&= \frac{1}{2}\left(-(a_{R_1} + ia_{I_1})\left(\frac{\partial \rho}{\partial x} + i\frac{\partial \iota}{\partial x}\right) - (a_{R_2} + ia_{I_2})\left(\frac{\partial^2 \rho}{\partial x^2} + i\frac{\partial^2 \iota}{\partial x^2}\right)\right. \\
&\quad -(a_{R_4} + ia_{I_4})(\rho + i\iota) + (a_{R_5} + ia_{I_5})|A|^2(\rho + i\iota) \\
&\quad +\delta(x - x_a)(u_R + iu_I) \\
&\quad -(a_{R_1} - ia_{I_1})\left(\frac{\partial \rho}{\partial x} - i\frac{\partial \iota}{\partial x}\right) - (a_{R_2} - ia_{I_2})\left(\frac{\partial^2 \rho}{\partial x^2} - i\frac{\partial^2 \iota}{\partial x^2}\right) \\
&\quad -(a_{R_4} - ia_{I_4})(\rho - i\iota) + (a_{R_5} - ia_{I_5})|A|^2(\rho - i\iota) \\
&\quad \left.+\delta(x - x_a)(u_R - iu_I)\right) \\
&= -a_{R_1}(x)\frac{\partial \rho}{\partial x} + a_{I_1}(x)\frac{\partial \iota}{\partial x} - a_{R_2}\frac{\partial^2 \rho}{\partial x^2} + a_{I_2}\frac{\partial^2 \iota}{\partial x^2} - a_{R_4}(x)\rho \\
&\quad +a_{I_4}(x)\iota + |A|^2 a_{R_5}\rho - |A|^2 a_{I_5}\iota + \delta(x - x_a)u_R \quad (4.228)
\end{aligned}$$

and

$$\frac{\partial \iota}{\partial t} = \frac{1}{2i}\left(\frac{\partial A}{\partial t} - \frac{\partial \bar{A}}{\partial t}\right)$$

$$= \frac{1}{2i}\left(-\left(a_{R_1} + ia_{I_1}\right)\left(\frac{\partial \rho}{\partial x} + i\frac{\partial \iota}{\partial x}\right) - \left(a_{R_2} + ia_{I_2}\right)\left(\frac{\partial^2 \rho}{\partial x^2} + i\frac{\partial^2 \iota}{\partial x^2}\right)\right.$$

$$- \left(a_{R_4} + ia_{I_4}\right)\left(\rho + i\iota\right) + \left(a_{R_5} + ia_{I_5}\right)|A|^2\left(\rho + i\iota\right)$$

$$+ \delta\left(x - x_a\right)\left(u_R + iu_I\right) \tag{4.229}$$

$$+ \left(a_{R_1} - ia_{I_1}\right)\left(\frac{\partial \rho}{\partial x} - i\frac{\partial \iota}{\partial x}\right) + \left(a_{R_2} - ia_{I_2}\right)\left(\frac{\partial^2 \rho}{\partial x^2} - i\frac{\partial^2 \iota}{\partial x^2}\right)$$

$$+ \left(a_{R_4} - ia_{I_4}\right)\left(\rho - i\iota\right) - \left(a_{R_5} - ia_{I_5}\right)|A|^2\left(\rho - i\iota\right) \tag{4.230}$$

$$\left. - \delta\left(x - x_a\right)\left(u_R - iu_I\right)\right)$$

$$= -a_{R_1}\left(x\right)\frac{\partial \iota}{\partial x} - a_{I_1}\left(x\right)\frac{\partial \rho}{\partial x} - a_{R_2}\frac{\partial^2 \iota}{\partial x^2} - a_{I_2}\frac{\partial^2 \rho}{\partial x^2} - a_{R_4}\left(x\right)\iota$$

$$- a_{I_4}\left(x\right)\rho + |A|^2 a_{R_5}\iota + |A|^2 a_{I_5}\rho + \delta\left(x - x_a\right)u_I. \tag{4.231}$$

Rearranging the terms, the equations become

$$\frac{\partial \rho}{\partial t} = -\left(a_{R_1}\left(x\right)\frac{\partial}{\partial x} + a_{R_2}\frac{\partial^2}{\partial x^2} + a_{R_4}\left(x\right) - a_{R_5}\left(\rho^2 + \iota^2\right)\right)\rho$$

$$+ \left(a_{I_1}\left(x\right)\frac{\partial}{\partial x} + a_{I_2}\frac{\partial^2}{\partial x^2} + a_{I_4}\left(x\right) - a_{I_5}\left(\rho^2 + \iota^2\right)\right)\iota + \delta\left(x - x_a\right)u_R \tag{4.232}$$

$$\frac{\partial \iota}{\partial t} = -\left(a_{I_1}\left(x\right)\frac{\partial}{\partial x} + a_{I_2}\frac{\partial^2}{\partial x^2} + a_{I_4}\left(x\right) - a_{I_5}\left(\rho^2 + \iota^2\right)\right)\rho$$

$$- \left(a_{R_1}\left(x\right)\frac{\partial}{\partial x} + a_{R_2}\frac{\partial^2}{\partial x^2} + a_{R_4}\left(x\right) - a_{R_5}\left(\rho^2 + \iota^2\right)\right)\iota + \delta\left(x - x_a\right)u_I \tag{4.233}$$

with boundary conditions $\rho\left(x \to \pm\infty, t\right) = 0$ and $\iota\left(x \to \pm\infty, t\right) = 0$.

Equations (4.232)–(4.233), with numerical values as given in [117, Appendix A] (reproduced in Appendix A), have been found to model well the dynamics of vortex shedding from a circular cylinder at Reynolds numbers near the critical Reynolds number, Re_c. Our objective is to design a state feedback controller that stabilizes the equilibrium $(\rho, \iota) = 0$ of (4.232)–(4.233). Figure 4.33 shows a sketch of the control system, superimposed on a visualization of vortex shedding. Based on the numerical values given in [117, Appendix A] (reproduced in Appendix A), we state the following assumptions which are assumed to hold throughout the analysis that follows.

Assumption 4.1 $a_2 \in (-\infty, 0)$ and $\mathfrak{Re}\left(a_5\right) \in (0, \infty)$, that is, $a_{R_2} < 0$, $a_{I_2} = 0$, and $a_{R_5} > 0$.

Uniform incoming flow

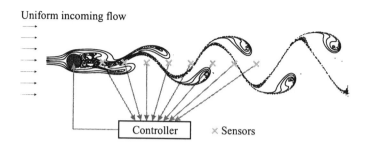

Figure 4.33: Vortex shedding from a cylinder visualized by passive tracer particles. The figure also shows the proposed control system configuration for suppression of vortex shedding [4].

Assumption 4.2 *For $|x| \gg 0$,*

$$\frac{1}{4a_{R_2}} a_{I_1}^2 (x) + a_{R_4} (x) \sim x^2. \tag{4.234}$$

The basic idea of the control design is to divide the domain into three separate parts; the upstream subsystem, the core, and the downstream subsystem, for which the following two facts are shown:

1. The upstream and downstream subsystems are input-to-state stable (in L_2 norm) with respect to certain boundary input terms.

2. A finite-difference approximation of any order of the core can be stabilized by state feedback, driving all the states to zero, including the boundary input terms of the upstream and downstream subsystems.

These two facts are treated in detail in Sections 4.4.3 and 4.4.4, respectively.

4.4.3   Energy Analysis

Lemma 4.5 *There exist real constants $x_u < 0$ and $x_d > 0$ such that solutions of system (4.232)–(4.233) satisfy*

$$\frac{d}{dt} \left(\|(\rho, \iota)\|_{L_2(-\infty, x_u)}^2 \right) \leq -c \, \|(\rho, \iota)\|_{L_2(-\infty, x_u)}^2$$
$$- \left(a_{R_1} (x) \left(\rho^2 + \iota^2 \right) + 2a_{R_2} \left(\frac{\partial \rho}{\partial x} \rho + \frac{\partial \iota}{\partial x} \iota \right) \right) \Bigg|_{x=x_u} \tag{4.235}$$

$$\frac{d}{dt}\left(\|(\rho, \iota)\|^2_{L_2(x_s, \infty)}\right) \leq -c\,\|(\rho, \iota)\|^2_{L_2(x_d, \infty)}$$
$$+\left(a_{R_1}(x)\left(\rho^2+\iota^2\right)+2a_{R_2}\left(\frac{\partial\rho}{\partial x}\rho+\frac{\partial\iota}{\partial x}\iota\right)\right)\Bigg|_{x=x_d} \qquad (4.236)$$

for some positive constant c.

Proof. The time derivative of $\|(\rho, \iota)\|^2_{L_2(a,b)}$ along solutions of (4.232)–(4.233) is

$$\frac{d}{dt}\left(\|(\rho, \iota)\|^2_{L_2(a,b)}\right) = 2\int_a^b \left(\rho\dot{\rho}+\iota\dot{\iota}\right) dx$$

$$= 2\int_a^b \Bigg[-\left(a_{R_1}(x)\frac{\partial\rho}{\partial x}+a_{R_2}\frac{\partial^2\rho}{\partial x^2}+a_{R_4}(x)\rho-|A|^2 a_{R_5}\rho\right)\rho$$
$$+\left(a_{I_1}(x)\frac{\partial\iota}{\partial x}+a_{I_2}\frac{\partial^2\iota}{\partial x^2}\right)\rho-\left(a_{I_1}(x)\frac{\partial\rho}{\partial x}+a_{I_2}\frac{\partial^2\rho}{\partial x^2}\right)\iota$$
$$-\left(a_{R_1}(x)\frac{\partial\iota}{\partial x}+a_{R_2}\frac{\partial^2\iota}{\partial x^2}+a_{R_4}(x)\iota-|A|^2 a_{R_5}\iota\right)\iota\Bigg]\,dx, \qquad (4.237)$$

where $a < b$ are arbitrary constants. Integration by parts yields

$$-2\int_a^b a_{R_1}(x)\frac{\partial\rho}{\partial x}\rho\,dx = -\left[a_{R_1}(x)\rho^2\right]_a^b+\int_a^b \frac{\partial a_{R_1}}{\partial x}\rho^2\,dx \qquad (4.238)$$

$$-2\int_a^b a_{R_1}(x)\frac{\partial\iota}{\partial x}\iota\,dx = -\left[a_{R_1}(x)\iota^2\right]_a^b+\int_a^b \frac{\partial a_{R_1}}{\partial x}\iota^2\,dx \qquad (4.239)$$

$$-2\int_a^b a_{R_2}\frac{\partial^2\rho}{\partial x^2}\rho\,dx = -2\left[a_{R_2}\frac{\partial\rho}{\partial x}\rho\right]_a^b+2\int_a^b a_{R_2}\left(\frac{\partial\rho}{\partial x}\right)^2\,dx \qquad (4.240)$$

$$-2\int_a^b a_{R_2}\frac{\partial^2\iota}{\partial x^2}\iota\,dx = -2\left[a_{R_2}\frac{\partial\iota}{\partial x}\iota\right]_a^b+2\int_a^b a_{R_2}\left(\frac{\partial\iota}{\partial x}\right)^2\,dx. \qquad (4.241)$$

Inserting (4.238)–(4.241) into (4.237), keeping in mind that $a_{I_2} = 0$ and $a_{R_5} < 0$ (Assumption 4.1), yields

$$\frac{d}{dt}\left(\|(\rho, \iota)\|^2_{L_2(a,b)}\right) \leq \left[-a_{R_1}(x)\left(\rho^2+\iota^2\right)-2a_{R_2}\left(\frac{\partial\rho}{\partial x}\rho+\frac{\partial\iota}{\partial x}\iota\right)\right]_a^b$$

$$+2\int_a^b \left[a_{R_2}\left(\frac{\partial\iota}{\partial x}\right)^2+a_{I_1}(x)\frac{\partial\iota}{\partial x}\rho-\left(-\frac{\partial a_{R_1}}{\partial x}+a_{R_4}(x)\right)\rho^2\right.$$

$$+a_{R_2}\left(\frac{\partial\rho}{\partial x}\right)^2 - a_{I_1}(x)\frac{\partial\rho}{\partial x}\iota - \left(-\frac{\partial a_{R_1}}{\partial x} + a_{R_4}(x)\right)\iota^2\Bigg]\,dx. \quad (4.242)$$

Now, consider the integrand in (4.242). We have that

$$a_{R_2}\left(\frac{\partial\iota}{\partial x}\right)^2 + a_{I_1}(x)\frac{\partial\iota}{\partial x}\rho - \left(-\frac{\partial a_{R_1}}{\partial x} + a_{R_4}(x)\right)\rho^2$$

$$+a_{R_2}\left(\frac{\partial\rho}{\partial x}\right)^2 - a_{I_1}(x)\frac{\partial\rho}{\partial x}\iota - \left(-\frac{\partial a_{R_1}}{\partial x} + a_{R_4}(x)\right)\iota^2$$

$$\leq \quad a_{R_2}\left|\frac{\partial\iota}{\partial x}\right|^2 + |a_{I_1}(x)|\left|\frac{\partial\iota}{\partial x}\right||\rho| - \left(-\frac{\partial a_{R_1}}{\partial x} + a_{R_4}(x)\right)\rho^2$$

$$+a_{R_2}\left|\frac{\partial\rho}{\partial x}\right|^2 + |a_{I_1}(x)|\left|\frac{\partial\rho}{\partial x}\right||\iota| - \left(-\frac{\partial a_{R_1}}{\partial x} + a_{R_4}(x)\right)\iota^2$$

$$= \quad a_{R_2}\left(\left|\frac{\partial\iota}{\partial x}\right| + \frac{1}{2a_{R_2}}|a_{I_1}(x)||\rho|\right)^2 - \left(\frac{1}{4a_{R_2}}a_{I_1}^2(x) - \frac{\partial a_{R_1}}{\partial x} + a_{R_4}(x)\right)\rho^2$$

$$a_{R_2}\left(\left|\frac{\partial\rho}{\partial x}\right| + \frac{1}{2a_{R_2}}|a_{I_1}(x)||\iota|\right)^2 - \left(\frac{1}{4a_{R_2}}a_{I_1}^2(x) - \frac{\partial a_{R_1}}{\partial x} + a_{R_4}(x)\right)\iota^2$$

$$\leq \quad -\left(\frac{1}{4a_{R_2}}a_{I_1}^2(x) - \frac{\partial a_{R_1}}{\partial x} + a_{R_4}(x)\right)(\rho^2 + \iota^2) \quad (4.243)$$

where we have used the fact that $a_{R_2} < 0$ (Assumption 4.1). Inserting (4.243) into (4.242) we obtain

$$\frac{d}{dt}\left(\|(\rho,\iota)\|_{L_2(a,b)}^2\right) \leq \left[-a_{R_1}(x)(\rho^2+\iota^2) - 2a_{R_2}\left(\frac{\partial\rho}{\partial x}\rho + \frac{\partial\iota}{\partial x}\iota\right)\right]_a^b$$

$$- 2\int_a^b \left(\frac{1}{4a_{R_2}}a_{I_1}^2(x) - \frac{\partial a_{R_1}}{\partial x} + a_{R_4}(x)\right)(\rho^2+\iota^2)\,dx. \quad (4.244)$$

By Assumption 4.2 there exist positive constants $x_u < 0$ and $x_d > 0$ such that

$$\frac{1}{4a_{R_2}}a_{I_1}^2(x) - \frac{\partial a_{R_1}}{\partial x} + a_{R_4}(x) > 0, \text{ for } x \in (-\infty, x_u) \cup (x_d, \infty), \quad (4.245)$$

and consequently, there exists a positive constant c, such that

$$\left(\frac{1}{4a_{R_2}}a_{I_1}^2(x) - \frac{\partial a_{R_1}}{\partial x} + a_{R_4}(x)\right) > \frac{1}{2}c, \text{ for } x \in (-\infty, x_u) \cup (x_d, \infty). \quad (4.246)$$

Inserting (4.246) into (4.244), (4.235) and (4.236) now follow by picking $(a, b) = (-\infty, x_u)$ and $(a, b) = (x_d, \infty)$, respectively, and applying the boundary conditions at $x = \pm\infty$. ∎

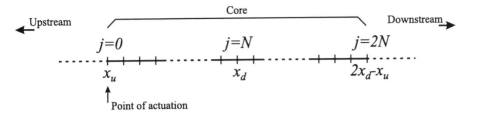

Figure 4.34: The system is discretized only in the core, which is the interval $[x_u, 2x_d - x_u]$, using a uniform grid with cell size h [4].

4.4.4 Stabilization by State Feedback

Discretization

Having determined the two constants x_u and x_d, which exist by Lemma 4.5, we discretize (4.232)–(4.233) using finite difference approximations on the interval $[x_u, 2x_d - x_u]$ as shown schematically in Figure 4.34. We define the finite difference approximations

$$\frac{\partial \rho}{\partial x}(jh) \approx \frac{\rho((j+1)h) - \rho(jh)}{h} \triangleq \frac{1}{h}\rho_{j+1} - \frac{1}{h}\rho_j \qquad (4.247)$$

$$\frac{\partial^2 \rho}{\partial x^2} \approx \frac{\rho((j+1)h) - 2\rho(jh) + \rho((j-1)h)}{h^2}$$

$$\triangleq \frac{\rho_{j+1}}{h^2} - 2\frac{\rho_j}{h^2} + \frac{\rho_{j-1}}{h^2} \qquad (4.248)$$

$$\frac{\partial \iota}{\partial x}(jh) \approx \frac{\iota((j+1)h) - \iota(jh)}{h} \triangleq \frac{1}{h}\iota_{j+1} - \frac{1}{h}\iota_j \qquad (4.249)$$

$$\frac{\partial^2 \iota}{\partial x^2} \approx \frac{\iota((j+1)h) - 2\iota(jh) + \iota((j-1)h)}{h^2}$$

$$\triangleq \frac{\iota_{j+1}}{h^2} - 2\frac{\iota_j}{h^2} + \frac{\iota_{j-1}}{h^2} \qquad (4.250)$$

where h is the grid cell size, and to simplify notation we set

$$r_{1,j} = a_{R_1}(x_j) \qquad (4.251)$$

$$i_{1,j} = a_{I_1}(x_j) \qquad (4.252)$$

$$r_2 = \frac{a_{R_2}}{h} \qquad (4.253)$$

$$r_{3,j} = ha_{R_4}(x_j) \qquad (4.254)$$

$$i_{3,j} = ha_{I_4}(x_j) \qquad (4.255)$$

$$r_4 = ha_{R_5} \qquad (4.256)$$

$$i_4 = ha_{I_5}. \qquad (4.257)$$

Inserting (4.247)–(4.257) into (4.232)–(4.233) we obtain the set of ordinary differential equations

$$h\dot{\rho}_j = -r_2\rho_{j-1} + \left(r_{1,j} + 2r_2 - r_{3,j} + r_4\left(\rho_j^2 + \iota_j^2\right)\right)\rho_j - \left(r_{1,j} + r_2\right)\rho_{j+1}$$
$$+ \left(-i_{1,j} + i_{3,j} - i_4\left(\rho_j^2 + \iota_j^2\right)\right)\iota_j + i_{1,j}\iota_{j+1} + \delta_{0,j}u_R \quad (4.258)$$

$$h\dot{\iota}_j = \left(i_{1,j} - i_{3,j} + i_4\left(\rho_j^2 + \iota_j^2\right)\right)\rho_j - i_{1,j}\rho_{j+1} - r_2\iota_{j-1}$$
$$+ \left(r_{1,j} + 2r_2 - r_{3,j} + r_4\left(\rho_j^2 + \iota_j^2\right)\right)\iota_j - \left(r_{1,j} + r_2\right)\iota_{j+1} + \delta_{0,j}u_I \quad (4.259)$$

for $j = 0, 1, 2, ..., 2N+2$, where $x_j \triangleq hj + x_u$ and $\delta_{i,j}$ denotes the Kronecker delta function. In (4.258)–(4.259) we have set the point of actuation to $x_a = x_u$, and used the fact that $a_{I_2} = 0$. Inserting (4.247)–(4.257) into (4.235) and (4.236) we obtain

$$\frac{d}{dt}\left(\|(\rho,\iota)\|_{L_2(-\infty,x_u)}^2\right) \le -c\,\|(\rho,\iota)\|_{L_2(-\infty,x_u)}^2$$
$$- r_{1,0}\left(\rho_0^2 + \iota_0^2\right) - 2r_2\left((\rho_1 - \rho_0)\rho_0 + (\iota_1 - \iota_0)\iota_0\right) \quad (4.260)$$

and

$$\frac{d}{dt}\left(\|(\rho,\iota)\|_{L_2(x_d,\infty)}^2\right) \le -c\,\|(\rho,\iota)\|_{L_2(x_d,\infty)}^2$$
$$+ r_{1,N}\left(\rho_N^2 + \iota_N^2\right) + 2r_2\left((\rho_{N+1} - \rho_N)\rho_N + (\iota_{N+1} - \iota_N)\iota_N\right), \quad (4.261)$$

which are the semi-discrete versions of (4.235)–(4.236).

Control design

The following Theorem summarizes our control design.

Theorem 4.2 *The control law defined recursively by the scheme*

$$\alpha_N \equiv \beta_N \equiv 0 \quad (4.262)$$

$$\alpha_{N-1}\left(\rho_N, \rho_{N+1}, \iota_{N+1}\right)$$
$$= \frac{1}{r_2}\left[\left(\frac{3}{2}r_{1,N} - r_{3,N} + c_N\right)\rho_N - r_{1,N}\rho_{N+1} + i_{1,N}\iota_{N+1}\right] \quad (4.263)$$

$$\beta_{N-1}\left(\rho_{N+1}, \iota_N, \iota_{N+1}\right)$$
$$= \frac{1}{r_2}\left[-i_{1,N}\rho_{N+1} + \left(\frac{3}{2}r_{1,N} - r_{3,N} + c_N\right)\iota_N - r_{1,N}\iota_{N+1}\right] \quad (4.264)$$

$$\alpha_{N-k}\left(\rho_j, \iota_j : j \in [N-k+1, N+k]\right) = \frac{1}{r_2}\left[c_{N-(k-1)}\left(\rho_{N-(k-1)} - \alpha_{N-(k-1)}\right)\right.$$

$$+ \left(r_{1,N-(k-1)} + 2r_2 - r_{3,N-(k-1)} + r_4\left(\rho^2_{N-(k-1)} + \iota^2_{N-(k-1)}\right)\right)\rho_{N-(k-1)}$$

$$- \left(r_{1,N-(k-1)} + 2r_2\right)\rho_{N-(k-2)}$$

$$+ \left(-i_{1,N-(k-1)} + i_{3,N-(k-1)} - i_4\left(\rho^2_{N-(k-1)} + \iota^2_{N-(k-1)}\right)\right)\iota_{N-(k-1)}$$

$$+ i_{1,N-(k-1)}\iota_{N-(k-2)} + r_2\alpha_{N-(k-2)}$$

$$- \sum_{j=N-k+2}^{N+k-1} \frac{\partial\alpha_{N-(k-1)}}{\partial\rho_j}\left(-r_2\rho_{j-1} + \left(r_{1,j} + 2r_2 - r_{3,j} + r_4\left(\rho^2_j + \iota^2_j\right)\right)\rho_j\right.$$

$$- \left(r_{1,j} + r_2\right)\rho_{j+1} + \left(-i_{1,j} + i_{3,j} - i_4\left(\rho^2_j + \iota^2_j\right)\right)\iota_j + i_{1,j}\iota_{j+1}\Big)$$

$$- \sum_{j=N-k+2}^{N+k-1} \frac{\partial\alpha_{N-(k-1)}}{\partial\iota_j}\left(\left(i_{1,j} - i_{3,j} + i_4\left(\rho^2_j + \iota^2_j\right)\right)\rho_j - i_{1,j}\rho_{j+1} - r_2\iota_{j-1}\right.$$

$$+ \left.\left(r_{1,j} + 2r_2 - r_{3,j} + r_4\left(\rho^2_j + \iota^2_j\right)\right)\iota_j - \left(r_{1,j} + r_2\right)\iota_{j+1}\right] \quad (4.265)$$

$$\beta_{N-k}\left(\rho_j, \iota_j : j \in [N-k+1, N+k]\right) = \frac{1}{r_2}\left[c_{N-(k-1)}\left(\iota_{N-(k-1)} - \beta_{N-(k-1)}\right)\right.$$

$$+ \left(i_{1,N-(k-1)} - i_{3,N-(k-1)} + i_4\left(\rho^2_{N-(k-1)} + \iota^2_{N-(k-1)}\right)\right)\rho_{N-(k-1)}$$

$$- i_{1,N-(k-1)}\rho_{N-(k-2)}$$

$$+ \left(r_{1,N-(k-1)} + 2r_2 - r_{3,N-(k-1)} + r_4\left(\rho^2_{N-(k-1)} + \iota^2_{N-(k-1)}\right)\right)\iota_{N-(k-1)}$$

$$- \left(r_{1,N-(k-1)} + 2r_2\right)\iota_{N-(k-2)} + r_2\beta_{N-(k-2)}$$

$$- \sum_{j=N-k+2}^{N+k-1} \frac{\partial\beta_{N-(k-1)}}{\partial\rho_j}\left(-r_2\rho_{j-1} + \left(r_{1,j} + 2r_2 - r_{3,j} + r_4\left(\rho^2_j + \iota^2_j\right)\right)\rho_j\right.$$

$$- \left(r_{1,j} + r_2\right)\rho_{j+1} + \left(-i_{1,j} + i_{3,j} - i_4\left(\rho^2_j + \iota^2_j\right)\right)\iota_j + i_{1,j}\iota_{j+1}\Big)$$

$$- \sum_{j=N-k+2}^{N+k-1} \frac{\partial\beta_{N-(k-1)}}{\partial\iota_j}\left(\left(i_{1,j} - i_{3,j} + i_4\left(\rho^2_j + \iota^2_j\right)\right)\rho_j - i_{1,j}\rho_{j+1} - r_2\iota_{j-1}\right.$$

$$+ \left.\left(r_{1,j} + 2r_2 - r_{3,j} + r_4\left(\rho^2_j + \iota^2_j\right)\right)\iota_j - \left(r_{1,j} + r_2\right)\iota_{j+1}\right], \quad (4.266)$$

for $k = 2, 3, 4, ..., N$, *and*

$$u_R\left(\rho_j, \iota_j : j \in [-1, 2N+1]\right)$$

$$= -\left[c_0\left(\rho_0 - \alpha_0\right) - r_2\rho_{-1} + \left(r_{1,0} + 2r_2 - r_{3,0} + r_4\left(\rho^2_0 + \iota^2_0\right)\right)\rho_0\right.$$

$$- \left(r_{1,0} + 2r_2\right)\rho_1 + \left(-i_{1,0} + i_{3,0} - i_4\left(\rho^2_0 + \iota^2_0\right)\right)\iota_0 + i_{1,0}\iota_1 + r_2\alpha_1$$

$$- \sum_{j=1}^{2N} \frac{\partial\alpha_0}{\partial\rho_j}\left(-r_2\rho_{j-1} + \left(r_{1,j} + 2r_2 - r_{3,j} + r_4\left(\rho^2_j + \iota^2_j\right)\right)\rho_j\right.$$

$$- (r_{1,j} + r_2)\, \rho_{j+1} + \left(-i_{1,j} + i_{3,j} - i_4 \left(\rho_j^2 + \iota_j^2\right)\right) \iota_j + i_{1,j}\iota_{j+1}$$

$$- \sum_{j=1}^{2N} \frac{\partial \alpha_0}{\partial \iota_j} \left(\left(i_{1,j} - i_{3,j} + i_4 \left(\rho_j^2 + \iota_j^2\right)\right) \rho_j - i_{1,j}\rho_{j+1} - r_2\iota_{j-1}\right.$$

$$\left. + \left(r_{1,j} + 2r_2 - r_{3,j} + r_4 \left(\rho_j^2 + \iota_j^2\right)\right) \iota_j - (r_{1,j} + r_2)\,\iota_{j+1}\right)] \quad (4.267)$$

$$u_I \left(\rho_j, \iota_j : j \in [-1, 2N+1]\right)$$
$$= - \left[c_0 \left(\iota_0 - \beta_0\right) + \left(i_{1,0} - i_{3,0} + i_4 \left(\rho_0^2 + \iota_0^2\right)\right) \rho_0 - i_{1,0}\rho_1\right.$$
$$- r_2\iota_{-1} + \left(r_{1,0} + 2r_2 - r_{3,0} + r_4 \left(\rho_0^2 + \iota_0^2\right)\right) \iota_0 - (r_{1,0} + 2r_2)\,\iota_1 + r_2\beta_1$$

$$- \sum_{j=1}^{2N} \frac{\partial \beta_0}{\partial \rho_j} \left(-r_2\rho_{j-1} + \left(r_{1,j} + 2r_2 - r_{3,j} + r_4 \left(\rho_j^2 + \iota_j^2\right)\right) \rho_j\right.$$

$$- (r_{1,j} + r_2)\, \rho_{j+1} + \left(-i_{1,j} + i_{3,j} - i_4 \left(\rho_j^2 + \iota_j^2\right)\right) \iota_j + i_{1,j}\iota_{j+1}$$

$$- \sum_{j=1}^{2N} \frac{\partial \beta_0}{\partial \iota_j} \left(\left(i_{1,j} - i_{3,j} + i_4 \left(\rho_j^2 + \iota_j^2\right)\right) \rho_j - i_{1,j}\rho_{j+1} - r_2\iota_{j-1}\right.$$

$$\left. + \left(r_{1,j} + 2r_2 - r_{3,j} + r_4 \left(\rho_j^2 + \iota_j^2\right)\right) \iota_j - (r_{1,j} + r_2)\,\iota_{j+1}\right)] \quad (4.268)$$

renders $(\rho_0, \iota_0, \rho_1, \iota_1, ..., \rho_N, \iota_N) = 0$ globally asymptotically stable. Moreover, solutions of system (4.232)–(4.233) satisfy

$$\|(\rho, \iota)\|_{L_2(-\infty, x_u)}^2 \rightarrow 0 \tag{4.269}$$

$$\|(\rho, \iota)\|_{L_2(x_d, \infty)}^2 \rightarrow 0 \tag{4.270}$$

as $t \rightarrow \infty$.

Proof. Consider the Lyapunov function candidate

$$V = \frac{h}{2} \sum_{j=0}^{N} \left[(\rho_j - \alpha_j)^2 + (\iota_j - \beta_j)^2\right] + \frac{1}{2}\|(\rho, \iota)\|_{L_2(x_d, \infty)}^2. \tag{4.271}$$

The time derivative of (4.271) along solutions of the system is

$$\dot{V} = \sum_{j=0}^{N-1} \left[(\rho_j - \alpha_j)(h\dot{\rho}_j - h\dot{\alpha}_j) + (\iota_j - \beta_j)\left(h\dot{\iota}_j - h\dot{\beta}_j\right)\right]$$

$$+ (\rho_N h\dot{\rho}_N + \iota_N h\dot{\iota}_N) + \frac{1}{2}\frac{d}{dt}\left[\|(\rho, \iota)\|_{L_2(x_d, \infty)}^2\right], \tag{4.272}$$

where we have used the fact that $\alpha_N = \beta_N = 0$. Inserting (4.263)–(4.264) into (4.272), yields

$$
\begin{aligned}
\dot{V} \leq\ & \sum_{j=0}^{N-1} \left[(\rho_j - \alpha_j)(h\dot{\rho}_j - h\dot{\alpha}_j) + (\iota_j - \beta_j)\left(h\dot{\iota}_j - h\dot{\beta}_j\right) \right] \\
& -r_2\rho_N\left(\rho_{N-1} - \alpha_{N-1}\right) - r_2\iota_N\left(\iota_{N-1} - r_2\beta_{N-1}\right) \\
& -\left(c_N - r_2\right)\left(\rho_N^2 + \iota_N^2\right) \\
& +r_4\left(\rho_N^2 + \iota_N^2\right)^2 - \frac{c}{2}\|(\rho,\iota)\|_{L_2(x_d,\infty)}^2 \\
\leq\ & \sum_{j=0}^{N-1} \left[(\rho_j - \alpha_j)(h\dot{\rho}_j - h\dot{\alpha}_j) + (\iota_j - \beta_j)\left(h\dot{\iota}_j - h\dot{\beta}_j\right) \right] \\
& -r_2\rho_N\left(\rho_{N-1} - \alpha_{N-1}\right) - r_2\iota_N\left(\iota_{N-1} - \beta_{N-1}\right) - \left(c_N - r_2\right)\left(\rho_N^2 + \iota_N^2\right) \\
& -\frac{c}{2}\|(\rho,\iota)\|_{L_2(x_d,\infty)}^2 \tag{4.273}
\end{aligned}
$$

where we have used the fact that $r_4 < 0$ (Assumption 4.1) in the last step. Next, we insert (4.258)–(4.259) into (4.273), and obtain

$$
\begin{aligned}
\dot{V} \leq\ & (\rho_0 - \alpha_0)\left[-r_2\rho_{-1} + \left(r_{1,0} + 2r_2 - r_{3,0} + r_4\left(\rho_0^2 + \iota_0^2\right)\right)\rho_0 - \left(r_{1,0} + r_2\right)\rho_1 \right. \\
& \left. + \left(-i_{1,0} + i_{3,0} - i_4\left(\rho_0^2 + \iota_0^2\right)\right)\iota_0 + i_{1,0}\iota_1 + u_R - h\dot{\alpha}_0 \right] \\
& + (\iota_0 - \beta_0)\left[\left(i_{1,0} - i_{3,0} + i_4\left(\rho_0^2 + \iota_0^2\right)\right)\rho_0 - i_{1,0}\rho_1 - r_2\iota_{-1} \right. \\
& \left. + \left(r_{1,0} + 2r_2 - r_{3,0} + r_4\left(\rho_0^2 + \iota_0^2\right)\right)\iota_0 - \left(r_{1,0} + r_2\right)\iota_1 + u_I - h\dot{\beta}_0 \right] \\
& + \sum_{j=1}^{N-1} \left[(\rho_j - \alpha_j)\left(-r_2\rho_{j-1} + \left(r_{1,j} + 2r_2 - r_{3,j} + r_4\left(\rho_j^2 + \iota_j^2\right)\right)\rho_j \right.\right. \\
& \left. - \left(r_{1,j} + r_2\right)\rho_{j+1} + \left(-i_{1,j} + i_{3,j} - i_4\left(\rho_j^2 + \iota_j^2\right)\right)\iota_j + i_{1,j}\iota_{j+1} - h\dot{\alpha}_j\right) \\
& + (\iota_j - \beta_j)\left(\left(i_{1,j} - i_{3,j} + i_4\left(\rho_j^2 + \iota_j^2\right)\right)\rho_j - i_{1,j}\rho_{j+1} - r_2\iota_{j-1} \right. \\
& \left.\left. + \left(r_{1,j} + 2r_2 - r_{3,j} + r_4\left(\rho_j^2 + \iota_j^2\right)\right)\iota_j - \left(r_{1,j} + r_2\right)\iota_{j+1} - h\dot{\beta}_j\right) \right] \\
& - r_2\rho_N\left(\rho_{N-1} - \alpha_{N-1}\right) - r_2\iota_N\left(\iota_{N-1} - \beta_{N-1}\right) - \left(c_N - r_2\right)\left(\rho_N^2 + \iota_N^2\right) \\
& - \frac{c}{2}\|(\rho,\iota)\|_{L_2(x_d,\infty)}^2 . \tag{4.274}
\end{aligned}
$$

At this point, we observe that the two summations in (4.265) in fact equal the time derivative of the previous α multiplied by h, that is $h\dot{\alpha}_{N-(k-1)}$. Similarly, the two summations in (4.266) equal the time derivative of the previous β multiplied by h, that is $h\dot{\beta}_{N-(k-1)}$. Therefore, from (4.265)–(4.266) we have

$$
\begin{aligned}
h\dot{\alpha}_j =\ & c_j\left(\rho_j - \alpha_j\right) - r_2\alpha_{j-1} + \left(r_{1,j} + 2r_2 - r_{3,j} + r_4\left(\rho_j^2 + \iota_j^2\right)\right)\rho_j \\
& - \left(r_{1,j} + 2r_2\right)\rho_{j+1} + \left(-i_{1,j} + i_{3,j} - i_4\left(\rho_j^2 + \iota_j^2\right)\right)\iota_j \\
& + i_{1,j}\iota_{j+1} + r_2\alpha_{j+1}, \tag{4.275}
\end{aligned}
$$

$$h\dot{\beta}_j \;=\; c_j \left(\iota_j - \beta_j\right) - r_2 \beta_{j-1} + \left(i_{1,j} - i_{3,j} + i_4 \left(\rho_j^2 + \iota_j^2\right)\right)\rho_j$$
$$-i_{1,j}\rho_{j+1} + \left(r_{1,j} + 2r_2 - r_{3,j} + r_4 \left(\rho_j^2 + \iota_j^2\right)\right)\iota_j$$
$$-\left(r_{1,j} + 2r_2\right)\iota_{j+1} + r_2 \beta_{j+1}, \tag{4.276}$$

for $j = 1, ..., N-1$. Furthermore, the two summations in (4.267) equal $h\dot{\alpha}_0$, and the two summations in (4.268) equal $h\dot{\beta}_0$. Keeping this in mind, and inserting (4.275)–(4.276) and (4.267)–(4.268) into (4.274), we get

$$\begin{aligned}
\dot{V} \;\leq\;& \left(\rho_0 - \alpha_0\right)\left(-c_0 \left(\rho_0 - \alpha_0\right) + r_2 \rho_1 - r_2 \alpha_1\right) \\
&+ \left(\iota_0 - \beta_0\right)\left(-c_0 \left(\iota_0 - \beta_0\right) + r_2 \iota_1 - r_2 \beta_1\right) \\
&+ \sum_{j=1}^{N-1}\left[\left(\rho_j - \alpha_j\right)\left(-r_2 \rho_{j-1} - c_j \left(\rho_j - \alpha_j\right) + r_2 \alpha_{j-1} + r_2 \rho_{j+1} - r_2 \alpha_{j+1}\right)\right. \\
&+ \left.\left(\iota_j - \beta_j\right)\left(-r_2 \iota_{j-1} - c_j \left(\iota_j - \beta_j\right) + r_2 \beta_{j-1} + r_2 \iota_{j+1} - r_2 \beta_{j+1}\right)\right] \\
&- r_2 \rho_N \left(\rho_{N-1} - \alpha_{N-1}\right) - r_2 \iota_N \left(\iota_{N-1} - \beta_{N-1}\right) - \left(c_N - r_2\right)\left(\rho_N^2 + \iota_N^2\right) \\
&- \frac{c}{2}\left\|(\rho,\iota)\right\|_{L_2(x_d,\infty)}^2 . \tag{4.277}
\end{aligned}$$

After rearranging the terms, we get

$$\begin{aligned}
\dot{V} \;\leq\;& -\sum_{j=0}^{N-1} c_j \left(\left(\rho_j - \alpha_j\right)^2 + \left(\iota_j - \beta_j\right)^2\right) \\
&+ r_2 \left(\rho_0 - \alpha_0\right)\left(\rho_1 - \alpha_1\right) + r_2 \left(\iota_0 - \beta_0\right)\left(\iota_1 - \beta_1\right) \\
&+ \sum_{j=1}^{N-1}\left[\left(\rho_j - \alpha_j\right)\left(-r_2 \rho_{j-1} + r_2 \alpha_{j-1} + r_2 \rho_{j+1} - r_2 \alpha_{j+1}\right)\right. \\
&+ \left.\left(\iota_j - \beta_j\right)\left(-r_2 \iota_{j-1} + r_2 \beta_{j-1} + r_2 \iota_{j+1} - r_2 \beta_{j+1}\right)\right] \\
&- r_2 \rho_N \left(\rho_{N-1} - \alpha_{N-1}\right) - r_2 \iota_N \left(\iota_{N-1} - \beta_{N-1}\right) \\
&- \left(c_N - r_2\right)\left(\rho_N^2 + \iota_N^2\right) - \frac{c}{2}\left\|(\rho,\iota)\right\|_{L_2(x_d,\infty)}^2 . \tag{4.278}
\end{aligned}$$

By splitting up the summation and changing summation indices, we obtain

$$\begin{aligned}
\dot{V} \;\leq\;& -\sum_{j=0}^{N-1} c_j \left(\left(\rho_j - \alpha_j\right)^2 + \left(\iota_j - \beta_j\right)^2\right) \\
&+ r_2 \left(\rho_0 - \alpha_0\right)\left(\rho_1 - \alpha_1\right) + r_2 \left(\iota_0 - \beta_0\right)\left(\iota_1 - \beta_1\right) \\
&- r_2 \left(\rho_1 - \alpha_1\right)\left(\rho_0 - \alpha_0\right) - r_2 \sum_{j=1}^{N-2}\left(\rho_{j+1} - \alpha_{j+1}\right)\left(\rho_j - \alpha_j\right) \\
&+ r_2 \left(\rho_{N-1} - \alpha_{N-1}\right)\left(\rho_N - \alpha_N\right) + r_2 \sum_{j=1}^{N-2}\left(\rho_j - \alpha_j\right)\left(\rho_{j+1} - \alpha_{j+1}\right) \\
&- r_2 \left(\iota_1 - \beta_1\right)\left(\iota_0 - \beta_0\right) - r_2 \sum_{j=1}^{N-2}\left(\iota_{j+1} - \beta_{j+1}\right)\left(\iota_j - \beta_j\right)
\end{aligned}$$

$$+r_2 \left(\iota_{N-1} - \beta_{N-1} \right) \left(\iota_N - \beta_N \right) + r_2 \sum_{j=1}^{N-2} \left(\iota_j - \beta_j \right) \left(\iota_{j+1} - \beta_{j+1} \right)$$

$$-r_2 \rho_N \left(\rho_{N-1} - \alpha_{N-1} \right) - r_2 \iota_N \left(\iota_{N-1} - \beta_{N-1} \right)$$

$$- \left(c_N - r_2 \right) \left(\rho_N^2 + \iota_N^2 \right) - \frac{c}{2} \| (\rho, \iota) \|_{L_2(x_d, \infty)}^2 , \tag{4.279}$$

and after cancellation of terms, we have

$$\dot{V} \leq - \sum_{j=0}^{N-1} c_j \left(\left(\rho_j - \alpha_j \right)^2 + \left(\iota_j - \beta_j \right)^2 \right)$$

$$- \left(c_N - r_2 \right) \left(\rho_N^2 + \iota_N^2 \right) - \frac{c}{2} \| (\rho, \iota) \|_{L_2(x_d, \infty)}^2 . \tag{4.280}$$

In the last step we used the fact that $\alpha_N = \beta_N = 0$. It now follows from standard results [90], that the equilibrium point $(\rho_0, \iota_0, \rho_1, \iota_1, ..., \rho_N, \iota_N) = 0$ is globally asymptotically stable, and that

$$\| (\rho, \iota) \|_{L_2(x_d, \infty)}^2 \to 0 \text{ as } t \to \infty. \tag{4.281}$$

Having established that $\rho_0, \iota_0, \rho_1, \iota_1 \to 0$ as $t \to \infty$, it follows from (4.235) in Lemma 4.5, and its semi-discrete version (4.260), that

$$\| (\rho, \iota) \|_{L_2(-\infty, x_u)}^2 \to 0 \text{ as } t \to \infty. \tag{4.282}$$

∎

4.4.5 Simulation Study

In order to demonstrate the performance of our backstepping controller, we present a simulation example. We set the Reynolds number to $Re = 50$, and discretize (4.232)–(4.233) on the domain $x \in [-5, 15]$ using 21 nodes. Homogeneous Dirichlet boundary conditions are enforced at $x = -5$ and $x = 15$. Next, we plot the expression (4.246), that is

$$\frac{1}{4a_{R_2}} a_{I_1}^2 (x) - \frac{\partial a_{R_1}}{\partial x} + a_{R_4} (x), \tag{4.283}$$

on our chosen domain. The result is shown in Figure 4.35. By inspection of the graph, we pick $x_u = -1.32$ and $x_d = 3.85$. The nodes of the discretization, along with x_u and x_d, are plotted in Figure 4.36. As the Figure shows, applying Theorem 4.2 at this point requires six backstepping steps. Instead, we use a discretization that is coarser for the control design, removing every other node. The remaining nodes, which are sensors, are shown in Figure 4.36 as circles. Now, only three steps of backstepping are required. The controller is generated using the symbolic toolbox in MATLAB, and is too complicated to write here. Figure 4.37 shows the values of (ρ, ι), represented by a vector originating at

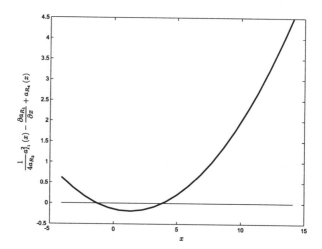

Figure 4.35: Graphical determination of the constants x_u and x_d for $R = 50$ using the numerical coefficients in [117, 4].

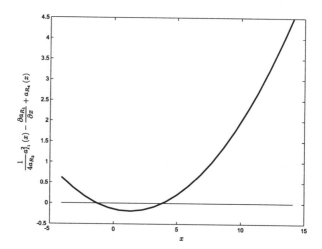

Figure 4.36: Locations of the grid points ($\times$), sensors ($\circ$), and $[x_u, x_d]$ ($+$). Only every second node is used for feedback, requiring three steps of backstepping in this case [4].

every node, for three time instances: $t = 200$, which is just before control is turned on, $t = 250$ and $t = 300$. As the Figure shows, the vectors alternate in direction and length indicating spatial unsteadiness reminiscent of vortex shedding. As time passes, the vectors become shorter, and eventually, all the states are driven to zero. Figure 4.38 shows the performance of the controller in terms of the output (ρ, ι) from one of the sensors (identified in Figure 4.36). As the Figure shows, the system is in the state of natural shedding for a couple of cycles, and then, at $t = 200$, the control is turned on. The measured state is effectively driven to zero.

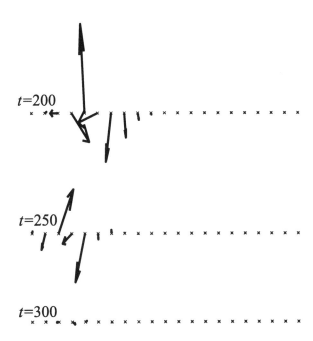

Figure 4.37: Nodal values of (ρ, ι) at three time instances: $t = 200$ (just before control is turned on), $t = 250$, and $t = 300$ [4].

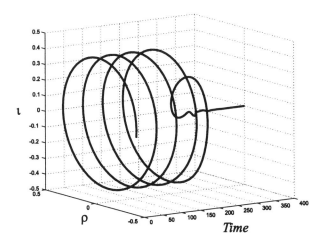

Figure 4.38: Time evolution of (ρ, ι) at sensor location number 4. Feedback is turned on at $t = 200$ [4].

Chapter 5

Mixing

A number of inherently different processes constitute what is called mixing. Ottino [110] distinguishes between three sub-problems of mixing: (i) mixing of a single fluid (or similar fluids) governed by the stretching and folding of material elements; (ii) mixing governed by diffusion or chemical reactions; and (iii) mixing of different fluids governed by the breakup and coalescence of material elements. Of course, all processes may be present simultaneously. In the first sub-problem, the interfaces between the fluids are passive [13], and the mixing may be determined by studying the movement of a passive tracer, or dye, in a homogeneous fluid flow. This is the problem we are interested in here.

In the following sections, we will review selected results on diagnostics of mixing based on dynamical systems theory, and then present new results on mixing enhancement using feedback control.

5.1 Dynamical Systems Approach

5.1.1 Chaotic Advection in the Blinking Vortex Flow

The application of dynamical systems theory to problems in mixing was initiated by Aref [12], who studied advection of passive tracer particles in the setting of an incompressible, inviscid fluid, contained in a 2D circular domain. It was shown in that reference, that a simple-looking, deterministic Eulerian velocity field may produce an essentially stochastic response in the Lagrangian advection characteristics of a passive tracer. This behavior is referred to as *chaotic advection*. The flow is driven by a point vortex, whose motion in a circular domain of radius a, is denoted $z(t)$, and it's strength is denoted Γ. The function $z(t)$ is referred to as the *stirring protocol*. A point in the domain is denoted ζ, so that $\zeta = x + iy$. The flow is represented by a complex valued function $f(\zeta)$, such that

$$w = f(\zeta) = \phi + i\psi$$

where ϕ is the potential function and ψ is the stream function. A point vortex at the origin with strength Γ is given by

$$w = \frac{\Gamma}{2\pi i} \ln \zeta.$$

If the point vortex is allowed to move according to $z(t)$, we get

$$w = \frac{\Gamma}{2\pi i} \ln (\zeta - z).$$

For a circular domain with radius a, we superpose the image of the point vortex at $z(t)$ to obtain

$$w = \frac{\Gamma}{2\pi i} \ln (\zeta - z) + \left(-\frac{\Gamma}{2\pi i} \ln \left(\zeta - \frac{a^2}{\bar{z}} \right) \right) = \frac{\Gamma}{2\pi i} \ln \left(\frac{\zeta - z}{\zeta - \frac{a^2}{\bar{z}}} \right).$$

Consider a particle p, placed into the domain. We denote it's position by

$$\zeta_p = x + iy.$$

The velocity of the particle is given by

$$\dot{\zeta}_p = u - iv = \frac{\partial \phi}{\partial x} + i\frac{\partial \psi}{\partial x} = \frac{\partial w}{\partial x} = \frac{\Gamma}{2\pi i} \left(\frac{1}{\zeta_p - z} - \frac{1}{\zeta_p - \frac{a^2}{\bar{z}}} \right). \qquad (5.1)$$

By solving (5.1) for an array of particles, we can now visualize the mixing properties of the flow for various stirring protocols. In [12], the particular case when

$$z(t) = \begin{cases} b & \text{for } nT \leq t < \left(n + \frac{1}{2}\right) T \\ -b & \text{for } \left(n + \frac{1}{2}\right) T \leq t < (n+1)T \end{cases} , \; n = 0, 1, 2, 3, \ldots \qquad (5.2)$$

is studied in detail. b is a constant (possibly complex) contained in the domain, and T is a positive real constant specifying the period of $z(t)$. The flow resulting from the stirring protocol (5.2) is called the blinking vortex flow. The evolution governed by (5.1) induces a mapping of the disk $|\zeta| \leq a$ onto itself, defined by $M : \zeta(t) \rightarrow \zeta(t + T)$. The mapping M is referred to as a Poincaré map. Denoting an initial set of points as P_0, and applying the map iteratively to obtain the sets

$$P_n = M(P_{n-1}), \; n = 1, 2, 3, \ldots$$

we define the set of visited points after N iterations as

$$\mathcal{P}_N = \bigcup_{0 \leq n \leq N} P_n.$$

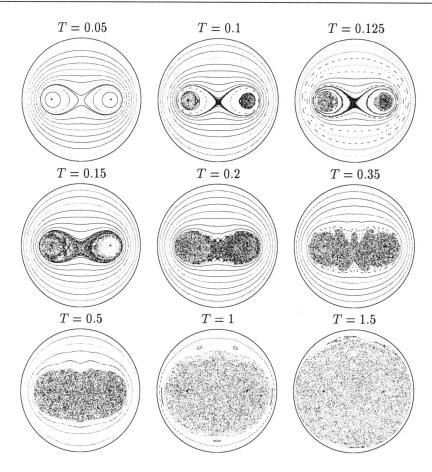

Figure 5.1: Iterated-map results ($\mathcal{P}_{2000}$) for $T \in [0.05, 0.1, 0.125, 0.15, 0.2, 0.35, 0.5, 1, 1.5]$ [12].

Figure 5.1 shows $\mathcal{P}_{2000}$ for $b = 0.5$, $P_0 = [-0.35, -0.2, -0.05, 0.1i, 0.2i, ..., 0.8i, 0.9i, 0.05, 0.2, 0.35]$, and for 9 different periods ($T \in [0.05, 0.1, 0.125, 0.15, 0.20, 0.35, 0.5, 1, 1.5]$). For $T = 0.05$, $\mathcal{P}_{2000}$ appears very regular, and the points visit a very small area of the total domain. This suggests that a blob of tracer material put into this flow, would be contained in a limited area given by the lines in the figure. As T is increased, regions of chaotic behavior appear, and the size of these regions increase with increasing T. When $T = 1.5$, no trace of regularity can be seen.

Next, we consider the kind of stirring experiment which would be carried out in a real device. This is done by introducing a blob of tracer material into the fluid at $t = 0$, and watch it evolve with time.

Figure 5.2 shows the initial configuration of particles representing a square blob

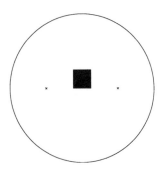

Figure 5.2: Initial configuration of particles for studying the mixing properties of the blinking vortex flow. The position of the point vortex is indicated by the two crosses ($\times$) [12].

for the study of the mixing properties of (5.2). 10000 particles are evenly distributed in the box $[-0.125, 0.125] \times [0, 0.25]$. The position of the point vortex is again chosen as $b = 0.5$. Figure 5.3 shows the configuration of particles for 4 different time instances ($t \in [3, 6, 9, 12]$) for 3 different periods ($T \in [0.1, 1, 3]$). As anticipated in Figure 5.1, the blob is contained in a very limited area for $T = 0.1$, and the shape of the area is easily recognized to be formed by the lines appearing in Figure 5.1. Thus, it is shown how the study of a dynamical system can be reduced to the study of a map, for which redundant dynamics are filtered out so that emphasis is put on the underlying structures that govern mixing and transport.

5.1.2 Particle Transport in the Mixing Region of the Oscillating Vortex Pair Flow

The Oscillating Vortex Pair Flow

In [116] dynamical systems theory is applied in a mathematically rigorous manner in order to study the mixing properties in a 2D model flow governed by a vortex pair in the presence of an oscillating external strain-rate field. The vortices have strength $\pm\Gamma$, and are initially separated by a distance $2d$ in the y direction. The stream function for the flow in a frame moving with the average velocity of the vortices is

$$\Psi = -\frac{\Gamma}{4\pi} log \left\{ \frac{(x - x_v)^2 + (y - y_v)^2}{(x - x_v)^2 + (y + y_v)^2} \right\} - V_v y + \epsilon x y sin(\omega t) ,$$

where $(x_v(t), y_v(t))$ and $(x_v(t), -y_v(t))$ are the vortex positions, ϵ is the strain rate and V_v is the average velocity of the vortex pair. For $\epsilon = 0$, $x_v(t) = 0$, $y_v(t) = d$ and $V_v = \Gamma/4\pi d$. Introducing appropriate scaling of the variables,

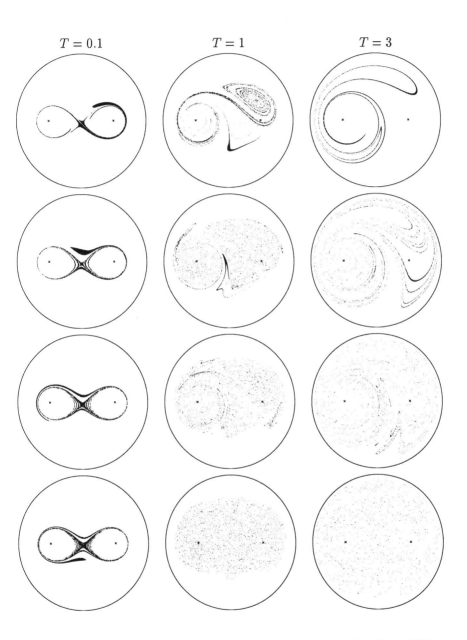

Figure 5.3: Configuration of particles in the blinking vortex flow for 4 different time instances ($t \in [3, 6, 9, 12]$) for 3 different periods ($T \in [0.1, 1, 3]$) [12].

and defining $\gamma = \Gamma/2\pi\omega d^2$ and $\nu_v = 2\pi dV_v/\Gamma$, we obtain the dimensionless equations for particle motion

$$\frac{dx}{dt} = -\left(\frac{y - y_v}{(x - x_v)^2 + (y - y_v)^2} - \frac{y + y_v}{(x - x_v)^2 + (y + y_v)^2}\right)$$
$$- \nu_v + \frac{\epsilon x}{\gamma} sin(t/\gamma) \quad (5.3)$$

$$\frac{dy}{dt} = (x - x_v)\left(\frac{1}{(x - x_v)^2 + (y - y_v)^2} - \frac{1}{(x - x_v)^2 + (y + y_v)^2}\right)$$
$$- \frac{\epsilon y}{\gamma} sin\,(t/\gamma). \quad (5.4)$$

The vortex positions are governed by

$$\frac{dx_v}{dt} = \frac{1}{2y_v} - \nu_v + \frac{\epsilon x_v}{\gamma} sin(t/\gamma)$$
$$\frac{dy_v}{dt} = -\frac{\epsilon y_v}{\gamma} sin\,(t/\gamma).$$

This flow approximates the motion in the vicinity of a vortex pair moving in a wavy-walled channel. In the perturbation analysis that follows, ϵ is assumed to be small. The right hand side of (5.3)–(5.4) can be expanded in powers of ϵ to obtain

$$\dot{x} = f_1(x, y) + \epsilon g_1(x, y, t/\gamma; \gamma) + O(\epsilon^2) \quad (5.5)$$
$$\dot{y} = f_2(x, y) + \epsilon g_2(x, y, t/\gamma; \gamma) + O(\epsilon^2) \quad (5.6)$$

where

$$f_1(x, y) = -\frac{y - 1}{I_-} + \frac{y + 1}{I_+} - \frac{1}{2}, \quad f_2(x, y) = x\left(\frac{1}{I_-} - \frac{1}{I_+}\right)$$

$$g_1(x, y, t/\gamma; \gamma) = (cos(t/\gamma) - 1)\left(\frac{1}{I_-} + \frac{1}{I_+} - \frac{2(y + 1)^2}{I_+^2}\right)$$
$$+ \frac{x}{\gamma} sin(t/\gamma)\left(\gamma^2\left(\frac{y - 1}{I_-^2} - \frac{y + 1}{I_+^2}\right) + 1\right) - \frac{1}{2}$$

$$g_2(x, y, t/\gamma; \gamma) = 2x(cos(t/\gamma) - 1)\left(\frac{y - 1}{I_-^2} + \frac{y + 1}{I_+^2}\right)$$
$$+ \frac{1}{\gamma} sin(t/\gamma)\left(\frac{\gamma^2}{2}\left(\frac{1}{I_-} - \frac{1}{I_+}\right) - x^2 y^2\left(\frac{1}{I_-^2} - \frac{1}{I_+^2}\right) - y\right)$$

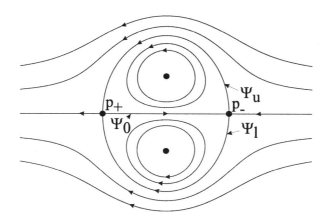

Figure 5.4: Streamlines of the unperturbed vortex pair flow [116].

$$I_+ = x^2 + (y+1)^2, \ I_- = x^2 + (y-1)^2.$$

The streamlines of the unperturbed flow ($\epsilon = 0$) are shown in Figure 5.4. For this case, there are two hyperbolic fixed points, p_- and p_+, connected by the limiting streamlines defined by

$$\Psi_l \ : \quad \Psi(x,y)|_{\epsilon=0} = 0, \ |x| < \sqrt{3}, \ y < 0$$
$$\Psi_0 \ : \quad \Psi(x,y)|_{\epsilon=0} = 0, \ |x| < \sqrt{3}, \ y = 0$$
$$\Psi_u \ : \quad \Psi(x,y)|_{\epsilon=0} = 0, \ |x| < \sqrt{3}, \ y > 0.$$

The particle motions on the interior of the limiting streamlines $\Psi_l \cup \Psi_u \cup p_- \cup p_+$ are qualitatively different from those on the exterior. Also, since streamlines cannot cross (for $\epsilon = 0$), there is no transport between the interior and the exterior of the limiting streamlines. We are interested in analyzing how this picture changes when the strain rate field is applied ($\epsilon \neq 0$).

The Poincaré Map

We may rewrite the time-varying system (5.5)–(5.6) as an equivalent time-invariant system by introducing the state $\theta = t/\gamma \mod 2\pi$. Thus, the system (5.5)–(5.6) can be written as

$$\dot{x} = f_1(x,y) + \epsilon g_1(x,y,\theta;\gamma) + O(\epsilon^2) \tag{5.7}$$
$$\dot{y} = f_2(x,y) + \epsilon g_2(x,y,\theta;\gamma) + O(\epsilon^2) \tag{5.8}$$
$$\dot{\theta} = 1/\gamma \tag{5.9}$$

which is now a three-dimensional time-invariant system. A two-dimensional global cross section of the three-dimensional state space of (5.7)–(5.9) can be defined by

$$\Sigma_{\tilde{\theta}} = \left\{ (x, y, \theta) | \theta = \tilde{\theta} \in [0, 2\pi) \right\}$$

on which we define the Poincaré map

$$T_{\tilde{\theta}} : \Sigma_{\tilde{\theta}} \rightarrow \Sigma_{\tilde{\theta}}$$
$$\left(x \left(\tilde{\theta} \right), y \left(\tilde{\theta} \right) \right) \mapsto \left(x \left(\tilde{\theta} + 2\pi \right), y \left(\tilde{\theta} + 2\pi \right) \right). \tag{5.10}$$

In the unperturbed case, the orbits of the Poincaré map are sequences of discrete points lying on the streamlines shown in Figure 5.4. Thus, the streamlines are invariant manifolds of the map. Orbits starting on Ψ_l, Ψ_0 and Ψ_u are heteroclinic orbits, and the points p_- and p_+ are fixed points of the map. Orbits on Ψ_l and Ψ_u approach p_+ in positive time, and $\Psi_l \cup \Psi_u \cup p_+$ is therefore the stable manifold of p_+, denoted W_+^s. Similarly, $\Psi_l \cup \Psi_u \cup p_-$ is the unstable manifold of p_-, denoted W_-^u. The unstable manifold of p_+, denoted W_+^u, is $\left\{ (x, y) | x < \sqrt{3}, y = 0 \right\}$, and the stable manifold of p_-, denoted W_-^s, is $\left\{ (x, y) | x > -\sqrt{3}, y = 0 \right\}$. Clearly, W_+^s and W_-^u intersect along Ψ_l and Ψ_u, creating a barrier to transport between the interior and exterior of the limiting streamlines. For sufficiently small ϵ, p_+ and p_- persist as fixed points of the Poincaré map (5.10), denoted $p_{+,\epsilon}$ and $p_{-,\epsilon}$, respectively. Their stable and unstable manifolds, W_+^s, W_-^s, W_+^u, and W_-^u also persist to become the stable and unstable manifolds of $p_{+,\epsilon}$ and $p_{-,\epsilon}$. They are denoted $W_{+,\epsilon}^s$, $W_{-,\epsilon}^s$, $W_{+,\epsilon}^u$, and $W_{-,\epsilon}^u$, respectively. Due to symmetry about the x-axis, $y = 0$ is an invariant manifold for all ϵ, which implies that the stable manifold of $p_{-,\epsilon}$ and the unstable manifold of $p_{+,\epsilon}$ always coincide on the line connecting the two fixed points. $W_{+,\epsilon}^s$ and $W_{-,\epsilon}^u$, on the other hand, may not coincide in the perturbed case. It is possible for $W_{+,\epsilon}^s$ and $W_{-,\epsilon}^u$ to intersect at an isolated point, which implies, by invariance of $W_{+,\epsilon}^s$ and $W_{-,\epsilon}^u$, that they must also intersect at every iterate of the Poincaré map and it's inverse. Thus, $W_{+,\epsilon}^s$ and $W_{-,\epsilon}^u$ may intersect at infinitely many discrete points, leading to a geometry like the one shown in Figure 5.5. If this is the case, the barrier that is present in the unperturbed case splits open, and transport of fluid across it becomes possible. This behavior of the stable and unstable manifolds is also reminiscent of chaotic particle motion.

Melnikov's Method

The existence of an isolated point of intersection of $W_{+,\epsilon}^s$ and $W_{-,\epsilon}^u$ can be established by Melnikov's method [133], which relates the signed distance between the two manifolds to the so-called Melnikov function according to

$$d(t_0, \epsilon) = \epsilon \frac{M(t_0)}{\|f(q_u(-t_0))\|} + O\left(\epsilon^2\right),$$

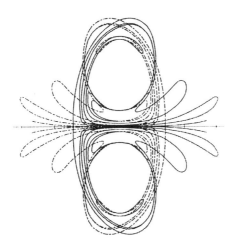

Figure 5.5: A Poincaré section of the oscillating vortex pair flow, showing the unstable (solid line) and stable (dashed line) manifolds of the two hyperbolic fixed points. Due to the tangling of the manifolds, this image is also referred to as the *homoclinic tangle* [116].

where $q_u(t)$ is the heteroclinic particle trajectory of the unperturbed velocity field, coinciding with Ψ_u of Figure 5.4, t_0 parametrizes distance along Ψ_u, and

$$\|f(q_u(-t_0))\| = \sqrt{f_1(q_u(-t_0))^2 + f_2(q_u(-t_0))^2}.$$

$M(t_0)$ is the Melnikov function defined as

$$M(t_0) = \int\limits_{-\infty}^{\infty} \{f_1(q_u(t))g_2(q_u(t), t + t_0) - f_2(q_u(t))g_1(q_u(t), t + t_0)\}\, dt.$$

The result of Melnikov states that simple zeros of $M(t_0)$ imply simple zeros of $d(t_0, \epsilon)$ (for sufficiently small ϵ). In [116], the Melnikov function for the system at hand is computed numerically to obtain

$$M(t_0) = \frac{F(\gamma)}{\gamma} sin(t_0/\gamma)$$

with $F(\gamma)$ plotted in Figure 5.6. For any fixed $\gamma \neq \gamma^*$, $M(t_0)$ has an infinite number of simple zeros, corresponding to transverse intersections of $W^s_{+,\epsilon}$ and $W^u_{-,\epsilon}$. This confirms the geometry shown in Figure 5.5. Studying the dynamics associated with the tangling of the stable and unstable manifolds of $p_{+,\epsilon}$ and $p_{-,\epsilon}$, can further quantify the particle transport taking place. This involves the motion of so-called *lobes*, and the area of these lobes quantifies transport. Again, for sufficiently small ϵ, the Melnikov function is a measure of the area of the lobes. We will not pursue this here, but refer the reader to [116] for further details.

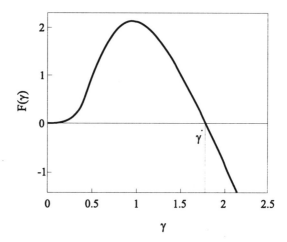

Figure 5.6: Graph of $F(\gamma)$ appearing in the Melnikov function [116].

5.1.3   Diagnostic Tools for Finite-Time Mixing

In time-periodic advection models, it is enough to know the velocity for finite times (e.g. for one period) to reproduce its infinite-time history. As a result, transport can be studied through the lobe dynamics of stable and unstable manifolds of appropriate Poincaré maps, as was demonstrated in the previous sections. However, for real flows, which in general are not periodic, one would have to know the velocity field for infinite times in order to define stable and unstable manifolds. This fact rules out the study of experimental datasets, both those consisting of measurements of actual flows, as well as those produced in a computer simulation. In a series of papers [62, 114, 115, 63, 64, 65], Haller and coworkers introduced and applied the notions of finite-time stable and unstable manifolds. The essentials of this new theory are outlined below.

Coherent Structures

A real flow will contain regions having different dynamical behavior. These regions are referred to as *Lagrangian coherent structures*. In Section 5.1.2 we encountered two fundamentally different dynamical behaviors in the oscillating vortex pair flow. One was the rotational motion occurring in the interior of the separating streamlines Ψ_l and Ψ_u, and the other was the translational motion occurring in the exterior of the separating streamlines. Thus, the regions in the interior and exterior of the separating streamlines are examples of coherent structures. The stable and unstable manifolds of the two fixed points coincided on the separating streamlines in the unperturbed case. In the perturbed case, however, they did not coincide, but intersected transversely at an infinite

Figure 5.7: Stretching across a coherent structure boundary [64].

Figure 5.8: Stretching along a coherent structure boundary leading to folding [64].

number of discrete points, leading to the formation of lobes whose dynamics is the mechanism by which transport between the interior and the exterior of the boundaries occurs. Such boundaries between coherent structures in the flow appear to be the locations in the flow where material blobs are stretched and folded most extensively. Extensive stretching and folding are reminiscent of effective mixing. Blobs of particles that travel together in the vicinity of coherent structure boundaries will in certain cases suddenly depart in opposite directions leading to local stretching *across* the boundary, as illustrated in Figure 5.7. In other cases the blob will become thinner and thinner as it is stretched *along* the coherent structure boundary leading, eventually, to folding due to the global geometry of the boundary, as illustrated in Figure 5.8. For these reasons, it is of interest to be able to localize the boundaries of coherent structures in a given finite-time dataset. A method that achieves this is presented next.

Material Lines and Surfaces

Consider the two-dimensional velocity field, $u(x, t)$, with the corresponding particle motion

$$\dot{x} = u(x, t) \tag{5.11}$$

on some *finite-time* interval $[t_{-1}, t_1]$. Given a curve of initial conditions, Γ_{t_0}, on the state space, later images of Γ_{t_0} under the motion (5.11), denoted Γ_t,

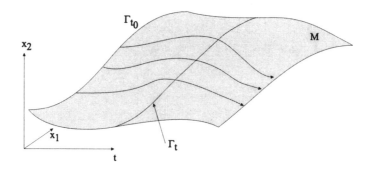

Figure 5.9: A material surface M spanned out by the material line Γ_t [64].

are called *material lines*. Augmenting the state space with the time variable, the evolving curve, Γ_t, spans a two-dimensional surface in the extended state space (x, t). This surface is called a *material surface*, denoted M, and sketched in Figure 5.9. In order to study the stability of M, we need to linearize the extended flow map, $\mathcal{F}^\tau$, along M. The extended flow map is given by

$$\mathcal{F}^\tau : \begin{pmatrix} x_0 \\ t_0 \end{pmatrix} \longmapsto \begin{pmatrix} x(t_0 + \tau; t_0, x_0) \\ t_0 + \tau \end{pmatrix}.$$

We denote the linearized extended flow map by $D\mathcal{F}^\tau$.

Stability Properties of Material Surfaces

M is called an unstable material surface on the time interval I_u if there is a positive exponent λ_u such that for any sufficiently close initial condition $p(\tau) = (x(\tau), \tau)$ and for any small time step $h > 0$, with $\tau \in I_u$ and $\tau + h \in I_u$, we have

$$dist(p(\tau + h), M) \geq dist(p(\tau), M) e^{\lambda_u h}.$$

So, if $\mathcal{N}(p, t)$ is a unit normal to M at the point (p, t) in the extended state space, then M is an unstable material surface over I_u if for all $\tau, \tau + h \in I_u$ and for all initial conditions $(x_0, \tau) \in M$, we have

$$\left| \mathcal{N}(x(\tau + h), t + h) \cdot D\mathcal{F}^h(x_0) \mathcal{N}(x_0, \tau) \right| \geq e^{\lambda_u h}, \tag{5.12}$$

where $(x(\tau + h), \tau + h)$ is the trajectory passing through (x_0, τ) in the extended state space. Figure 5.10 shows a sketch of an unstable material surface over the time interval I_u. N is called a stable material surface if it is an unstable material surface in the sense of (5.12) backwards in time. An unstable (respectively, stable) material line with instability interval I_u (respectively, stability interval I_s) is a curve Γ_t which generates an unstable (respectively, stable) material

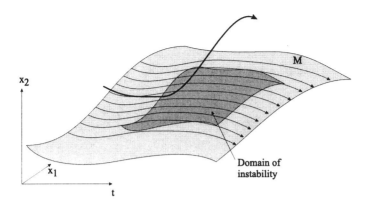

Figure 5.10: An unstable material surface repelling nearby trajectories [64].

surface in the extended state space. Unstable and stable material lines and surfaces are referred to as *hyperbolic material lines* and *hyperbolic material surfaces*, respectively. Their associated intervals I_u and I_s are referred to as *hyperbolicity intervals*. Hyperbolic material surfaces are never unique on a finite-time interval. However, if they are unstable for a sufficiently long time interval, they will appear unique up to numerically unresolvable errors. Thus, stable and unstable material surfaces will be used in the definition of coherent structure boundaries given below. The results outlined above have also been extended to the 3D setting [65].

Detecting Coherent Structure Boundaries

For any initial condition x_0 at time $t_0 \in [t_{-1}, t_1]$, consider the maximal open set, $\mathcal{I}_u(x_0)$, within $[t_0, t_1]$ on which the instability condition (5.12) is satisfied. The instability time $T_u(x_0, t_0)$ associated with x_0 over the time interval $[t_0, t_1]$ is defined as

$$T_u(x_0, t_0) = \frac{1}{t_1 - t_0} \int\limits_{\mathcal{I}_u(x_0)} dt.$$

That is, T_u is the fraction of the time $t_1 - t_0$, for which the instability condition (5.12) holds. Similarly, the maximal open set, $\mathcal{I}_s(x_0)$, within $[t_0, t_{-1}]$, on which the instability condition (5.12) is satisfied in backward time, defines the stability time $T_s(x_0, t_0)$ associated with x_0 over the backward time interval $[t_0, t_{-1}]$ as

$$T_s(x_0, t_0) = \frac{1}{t_0 - t_{-1}} \int\limits_{\mathcal{I}_s(x_0)} dt.$$

$T_u(x_0, t_0)$ and $T_s(x_0, t_0)$ are called the *hyperbolicity times* associated with x_0 at t_0. With these definitions, coherent structure boundaries at $t = t_0$ are given by stable and unstable material lines along which T_s or T_u attains local extrema. In the next section, the fields T_s and T_u are plotted for studying mixing properties in controlled 2D channel flow. It is interesting to note, that for time periodic velocity fields, the coherent structure boundaries defined above coincide with the stable and unstable manifolds of hyperbolic fixed points of the corresponding Poincaré map.

5.2 Destabilization of 2D Channel Flow

5.2.1 Numerical Simulations

Mixing is commonly induced by means of open loop methods such as mechanical stirring, jet injection or mixing valves. These methods may use excessive amounts of energy, which in certain cases is undesirable. In [1], it is proposed to use active feedback control on the boundary of a 2D channel flow, in order to exploit the natural tendency in the flow to mix. The results of Section 4.3.1 show that the control law (4.63)–(4.64) has a significant stabilizing influence on the 2D channel flow. In this section, we explore the behavior of the flow when k_v is chosen such that this feedback destabilizes the flow rather than stabilizes it. The conjecture is that the flow will develop a complicated pattern in which mixing will occur. 2D simulations are performed at $Re = 6000$, for which the parabolic equilibrium profile is unstable. The vorticity map for the fully established flow (uncontrolled) at this Reynolds number is shown in the topmost graph in Figure 5.11. This is the initial data for the simulations. Some mixing might be expected in this flow, as it periodically ejects vorticity into the core of the channel. The objective, however, is to enhance the mixing process by boundary control, which we impose by setting $k_v = 0.1$ in (4.64). The vorticity maps in Figure 5.11 suggest that the flow pattern becomes considerably more complicated as a result of the control. The upper-left and upper-right graphs in Figure 5.12 show the perturbation energy, $E(\mathbf{w})$, and enstrophy as functions of time. The former increases by a factor of 5, while the latter is doubled. It is interesting to notice that the control leading to such an agitated flow is small (see lower-left graph in Figure 5.12). The maximum value of the control flow kinetic energy is less than 0.7% of the perturbation kinetic energy of the uncontrolled flow, and only about 0.1% of the fully developed, mixed (controlled) flow! Next, we will quantify the mixing in a more rigorous way, by studying the movement of passive tracer particles, representing dye blobs.

The location of the dye as a function of time completely describes the mixing, but in a flow that mixes well, the length of the interface between the dye and the fluid increases exponentially with time. Thus, calculating the location of the dye for large times is not feasible within the restrictions of modest computer resources [51]. We do, nevertheless, attempt this for small times, and supplement the results with less accurate, but computationally feasible, calculations

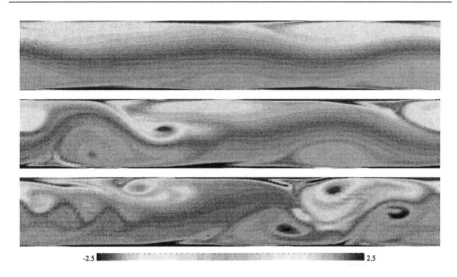

-2.5   2.5

Figure 5.11: Vorticity map for the fully established, uncontrolled, channel flow at $Re = 6000$ (top), and for the controlled case at $t = 50$ (middle) and $t = 80$ (bottom) [1].

for larger times. A particle-line method, loosely based on [129] and [88], is used to track the dye interface. In short, this method represents the interface as a number of particles connected by straight lines. The positions of the particles are governed by the equation $dX/dt = (U(X,t), V(X,t))$, where X is a vector of particle positions. At the beginning of each time step, new particles are added such that at the end of the time step, a prescribed resolution, given in terms of the maximum length between neighboring particles, is maintained. The fact that we are working with a single fluid representing multiple miscible fluids, ensures that dye surfaces remain connected [111]. At $t = 50$, when the perturbation energy is about tripled in the controlled case (Figure 5.12), eighteen blobs are distributed along the centerline of the channel as shown in Figure 5.13. They cover 25% of the total domain.

Figure 5.14 compares the configuration of the dye in the uncontrolled and controlled cases for 5 time instances. The difference in complexity is clear, however, large regions are poorly mixed even at $t = 85$. The lower-right graph in Figure 5.12 shows the total length of the surface of the dye. The length appears to grow linearly with time in the uncontrolled case, whereas for the controlled case, it grows much faster, reaching values an order of magnitude larger than in the uncontrolled case. In order to approximate the dye distribution for large time, a fixed number of particles are uniformly distributed throughout the domain, distinguishing between particles placed on the inside (*black* particles) and on the outside (*white* particles) of regions occupied by dye. Figure 5.15 shows the distribution of black particles at $t = 85$ (for comparison with Figure 5.14), 100,

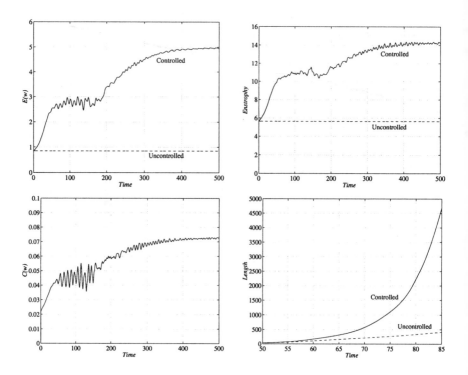

Figure 5.12: Energy $E(\mathbf{w})$ (top left), enstropy (top right), control effort $C(\mathbf{w})$ (bottom left), and dye surface length (bottom right), as functions of time [1].

Figure 5.13: Initial distribution of dye blobs (at $t = 50$) [1].

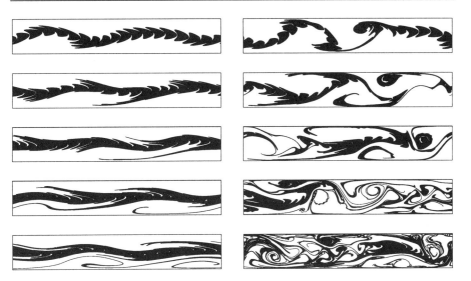

Figure 5.14: Dye distribution for uncontrolled flow (left column) versus controlled flow (right column) at $t = 55$, 60, 65, 75 and 85 (from top towards bottom) [1].

125 and 150. The particle distribution becomes increasingly uniform.

In order to quantify the mixing further, we ask the following question: given a box of size ε, what is the probability, P, of the fluid inside being *well mixed*? An appropriate choice of ε, and what is considered well mixed, are application specific parameters, and are usually given by requirements of some downstream process. In our case, the blobs initially cover 25% of the domain, so we will define *well mixed* to mean that the dye covers between 20% and 30% of the area of the box. The size ε of the boxes will be given in terms of pixels along one side of the box, so that the box covers ε^2 pixels out of a total of 2415×419 pixels for the entire domain. On this canvas, the box may be placed in $(419 - (\varepsilon - 1)) \times 2415$ different locations. The fraction of area covered by dye inside box i of size ε, is for small times calculated according to

$$c_\varepsilon^i = \frac{n_p}{\varepsilon^2} \qquad (5.13)$$

where n_p is the number of pixels covered by dye, and for large times according to

$$c_\varepsilon^i = \frac{n_b}{n_w + n_b} \qquad (5.14)$$

where n_b and n_w denote the number of black and white particles, respectively,

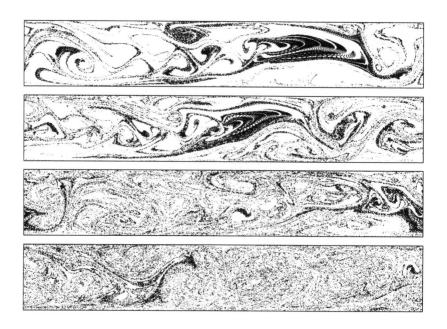

Figure 5.15: Particle distribution for controlled flow at $t = 85, 100, 125$ and 150 (from top towards bottom) [1].

contained in the box. P , which depends on ε, is calculated as follows

$$P_\varepsilon = \frac{1}{n} \sum_{i=1}^{n} eval(0.2 < c_\varepsilon^i < 0.3) \qquad (5.15)$$

where n is the total number of boxes. The expression in the summation evaluates to 1 when $0.2 < c_\varepsilon^i < 0.3$ and 0 otherwise. For small times $n = (419 - (\varepsilon - 1)) \times 2415$, whereas for large times n may be smaller as we choose to ignore boxes containing less than 25 particles. Figures 5.16 and 5.17 show P_ε as a function of time for $\varepsilon \in [15, 30, 45, 60]$. For all cases, the probability of the contents of the box being well mixed increases with time. In Figure 5.18, the hyperbolicity times of a grid of uniformly distributed initial conditions at $t_0 = 100$ are shown (see Section 5.1.3 for the definition of hyperbolicity time). The geometry of the coherent structure boundaries become considerable more complex in the controlled case, indicating extensive stretching and folding of material elements in the flow.

In conclusion, we have achieved substantial mixing enhancement using relatively small control effort, by exploiting instability mechanisms inherent in the flow.

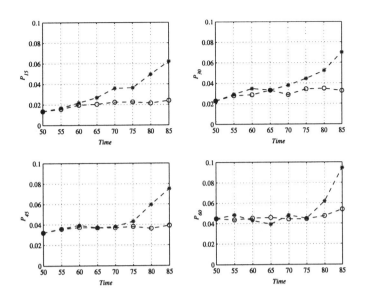

Figure 5.16: Probability of well mixedness for the uncontrolled case (o) and controlled case (*) [1].

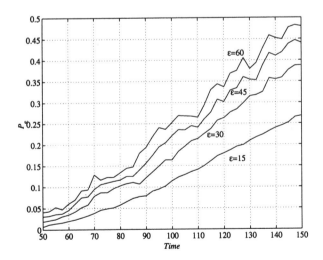

Figure 5.17: Probability of well mixedness for the controlled case based on uniform particle distribution [1].

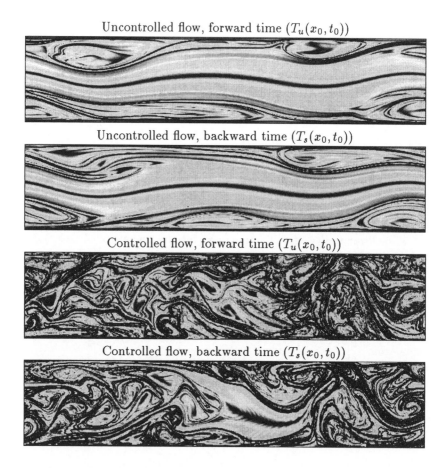

Figure 5.18: Hyperboliticy times for uncontrolled (upper two graphs) and controlled (lower two graphs) channel flow at $t_0 = 100$. The graphs were produced by Dr. Yong Wang, Brown University.

5.3 Optimal Mixing in 3D Pipe Flow

Recently, the results of the previous section have been generalized to 3D pipe flow [17] (also in [3]). Using a control law similar to that designed for the stabilizing case in (4.221), enhanced instability of the parabolic equilibrium profile is achieved, and the control law is shown to exhibit optimality properties. This will be made clear in the following derivations, originating from the Navier-Stokes equation for 3D pipe flow stated in (2.64) and (2.65)–(2.67).

5.3.1 Sensing and Actuation

As mentioned in the previous section, the boundary conditions on the wall of the pipe incorporate our actuation. The fluid velocity at the wall is restricted to be normal to the wall, that is, we take v_{r-w} as the control input, and set $v_{\theta-w} = v_{z-w} = 0$, where we have defined, for notational convenience, the variables on the wall as

$$v_{r-w}\left(\theta, z, t\right) \triangleq v_r\left(1, \theta, z, t\right), \; v_{\theta-w}\left(\theta, z, t\right) \triangleq v_\theta\left(1, \theta, z, t\right), \text{ and}$$
$$v_{z-w}\left(\theta, z, t\right) \triangleq v_z\left(1, \theta, z, t\right).$$

We also impose on the control input that it satisfies

$$v_{r-w}\left(\theta, z, t\right) = -v_{r-w}\left(\theta + \pi, z, t\right), \tag{5.16}$$

which states that if suction is applied at a point (θ, z) on the pipe wall, then an equal amount of blowing is applied at the opposite point $(\theta + \pi, z)$. This is illustrated in Figure 5.19. It is clear that condition (5.16) ensures a zero net mass flux across the pipe wall, and therefore it is a natural condition to impose from a mass balance point of view. The measurement available is the pressure drop, denoted Δp, from any point (θ, z) on the pipe wall to the opposite point $(\theta + \pi, z)$. That is,

$$\Delta p(\theta, z, t) \triangleq p\left(1, \theta, z, t\right) - p\left(1, \theta + \pi, z, t\right). \tag{5.17}$$

5.3.2 Measures of Mixing

There are two key ingredients to effective mixing. The fluid flow field must inflict extensive stretching to material elements, and the stretching should be accompanied by folding. In this work, we define two measures of the fluid flow field that are instrumental to our development below. One is the kinetic energy

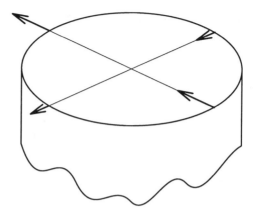

Figure 5.19: Actuation is symmetric about the pipe centerline [17].

of the perturbation, defined as

$$E\left(\mathbf{w}\right) \triangleq \lim_{\varepsilon \to 0} E_\varepsilon(\mathbf{w}) \triangleq \lim_{\varepsilon \to 0} \frac{1}{2} \int_0^L \int_0^{2\pi} \int_\varepsilon^1 \left(v_r^2 + v_\theta^2 + v_z^2\right) r \, dr \, d\theta \, dz, \qquad (5.18)$$

and the other is a measure of spatial velocity gradients, defined as

$$m\left(\mathbf{w}\right) \triangleq \int_0^L \int_0^{2\pi} \int_0^1 \left[\left(\frac{\partial v_r}{\partial z}\right)^2 + \left(\frac{\partial v_\theta}{\partial z}\right)^2 + \left(\frac{\partial v_z}{\partial r}\right)^2 + \left(\frac{1}{r}\frac{\partial v_z}{\partial \theta}\right)^2 + \left(\frac{\partial v_z}{\partial z}\right)^2 \right.$$

$$\left. + \left(\frac{\partial v_r}{\partial r}\right)^2 + \left(\frac{\partial v_\theta}{\partial r}\right)^2 + \left(\frac{v_r}{r} + \frac{1}{r}\frac{\partial v_\theta}{\partial \theta}\right)^2 + \left(\frac{v_\theta}{r} - \frac{1}{r}\frac{\partial v_r}{\partial \theta}\right)^2 \right] r \, dr \, d\theta \, dz. \quad (5.19)$$

The latter measure, (5.19), appears to be stronger connected to mixing. While it is clear that stretching of material elements is explicit in a measure of spatial gradients of the flow field, folding is implicit in the measure due to the boundedness of the flow domain, and the fact that $\mathbf{w}$ satisfies the Navier-Stokes equation. Thus, our objective becomes that of designing a feedback control law, in terms of suction and blowing of fluid normally to the pipe wall, that is optimal with respect to some meaningful cost functional related to $m(\mathbf{w})$.

5.3.3 Energy Analysis

Before giving the main result on controller design and optimality, we state two key lemmas that are needed frequently in what follows. The first lemma is a Lyapunov type result and it relates the time derivative of $E\left(\mathbf{w}\left(t\right)\right)$ to $m\left(\mathbf{w}\left(t\right)\right)$. The second lemma provides a bound on a crossterm in the streamwise (v_z) and radial (v_r) velocities, originating from the nonlinear convective terms in the Navier-Stokes equations.

Lemma 5.1 *If $v_{\theta-w}$ and v_{z-w} are zero, and v_{r-w} satisfies (5.16), then*

$$\dot{E}(\mathbf{w}) = -\frac{1}{2}\int_0^L\int_0^{2\pi} v_{r-w}\Delta p\,d\theta dz - \frac{1}{Re}\int_0^L\int_0^{2\pi} v_{r-w}^2\,d\theta dz$$

$$-\frac{1}{Re}m(\mathbf{w}) - \int_0^L\int_0^{2\pi}\int_0^1 rv_zv_r\frac{\partial\tilde{V}_z}{\partial r}dr d\theta dz \quad (5.20)$$

along solutions of system (2.64)–(2.67).

Proof. The time derivative of $E_\varepsilon(\mathbf{w})$ along trajectories of (2.64)–(2.67) was calculated in Section 4.3.4 and the result was

$$\dot{E}_\varepsilon(\mathbf{w}) = \int_0^L\int_0^{2\pi}\left[-\frac{1}{2}rv_r^3\right]_\varepsilon^1 d\theta dz + \int_0^L\int_0^{2\pi}\left[-rv_rp\right]_\varepsilon^1 d\theta dz + \frac{1}{Re}\int_0^L\int_0^{2\pi}\left[rv_r\frac{\partial v_r}{\partial r}\right]_\varepsilon^1 d\theta dz$$

$$+\frac{1}{2}\int_0^L\int_0^{2\pi}\left[-rv_rv_\theta^2\right]_\varepsilon^1 d\theta dz + \frac{1}{Re}\int_0^L\int_0^{2\pi}\left[rv_\theta\frac{\partial v_\theta}{\partial r}\right]_\varepsilon^1 d\theta dz + \frac{1}{2}\int_0^L\int_0^{2\pi}\left[-rv_z^2v_r\right]_\varepsilon^1 d\theta dz$$

$$+\frac{1}{Re}\int_0^L\int_0^{2\pi}\left[rv_z\frac{\partial v_z}{\partial r}\right]_\varepsilon^1 d\theta dz - \int_0^L\int_0^{2\pi}\int_\varepsilon^1 rv_zv_r\frac{\partial\tilde{V}_z}{\partial r}dr d\theta dz$$

$$-\frac{1}{Re}\int_0^L\int_0^{2\pi}\int_\varepsilon^1 r\left(\frac{\partial v_r}{\partial z}\right)^2 dr d\theta dz - \frac{1}{Re}\int_0^L\int_0^{2\pi}\int_\varepsilon^1 r\left(\frac{\partial v_\theta}{\partial z}\right)^2 dr d\theta dz$$

$$-\frac{1}{Re}\int_0^L\int_0^{2\pi}\int_\varepsilon^1 r\left(\frac{\partial v_z}{\partial r}\right)^2 dr d\theta dz - \frac{1}{Re}\int_0^L\int_0^{2\pi}\int_\varepsilon^1 \frac{1}{r}\left(\frac{\partial v_z}{\partial\theta}\right)^2 dr d\theta dz$$

$$-\frac{1}{Re}\int_0^L\int_0^{2\pi}\int_\varepsilon^1 r\left(\frac{\partial v_z}{\partial z}\right)^2 dr d\theta dz - \frac{1}{Re}\int_0^L\int_0^{2\pi}\int_\varepsilon^1 r\left(\frac{\partial v_r}{\partial r}\right)^2 dr d\theta dz$$

$$-\frac{1}{Re}\int_0^L\int_0^{2\pi}\int_\varepsilon^1 r\left(\frac{\partial v_\theta}{\partial r}\right)^2 dr d\theta dz - \frac{1}{Re}\int_0^L\int_0^{2\pi}\int_\varepsilon^1 \frac{1}{r}\left(v_r + \frac{\partial v_\theta}{\partial\theta}\right)^2 dr d\theta dz$$

$$-\frac{1}{Re}\int_0^L\int_0^{2\pi}\int_\varepsilon^1 \frac{1}{r}\left(v_\theta - \frac{\partial v_r}{\partial\theta}\right)^2 dr d\theta dz. \quad (5.21)$$

Since v_{r-w} has the form (5.16), we have for odd n

$$\int_0^{2\pi} v_{r-w}^n d\theta = \int_{-\pi}^0 v_{r-w}^n d\theta + \int_0^\pi v_{r-w}^n d\theta$$

$$= -\int_{-\pi}^{0} v_{r-w}^{n}\left(-\theta, z, t\right) d\theta + \int_{0}^{\pi} v_{r-w}^{n}\left(\theta, z, t\right) d\theta$$

and by a change of variables in the first integral $(\theta^* = -\theta)$, we get

$$\int_{0}^{2\pi} v_{r-w}^{n} d\theta = -\int_{0}^{\pi} v_{r-w}^{n}\left(\theta^*, z, t\right) d\theta^* + \int_{0}^{\pi} v_{r-w}^{n}\left(\theta, z, t\right) d\theta = 0.$$

This proves that the net mass flow through the wall is zero, and that

$$\int_{0}^{2\pi} v_{r-w}^{3} d\theta = 0 \tag{5.22}$$

by setting $n = 1$ and $n = 3$, respectively. Furthermore, we have

$$\int_{0}^{2\pi} [v_r p]_{r=1} \, d\theta = \int_{0}^{\pi} [v_r p]_{r=1} \, d\theta + \int_{\pi}^{2\pi} [v_r p]_{r=1} \, d\theta$$

$$= \int_{0}^{\pi} [v_r p]_{r=1} \, d\theta - \int_{0}^{\pi} v_{r-w}\left(\theta, z, t\right) p\left(1, \theta + \pi, z, t\right) d\theta$$

$$= \int_{0}^{\pi} v_{r-w}\left(\theta, z, t\right) \Delta p \, d\theta = \frac{1}{2}\int_{0}^{2\pi} v_{r-w}\left(\theta, z, t\right) \Delta p \, d\theta \tag{5.23}$$

and, since $v_{\theta-w}$ and v_{z-w} are zero, (2.64) yields

$$\frac{\partial v_r}{\partial r}\left(1, \theta, z, t\right) = -v_r\left(1, \theta, z, t\right). \tag{5.24}$$

By inserting $v_{\theta-w}\left(\theta, z, t\right) = v_{z-w}\left(\theta, z, t\right) = 0$ and (5.22)–(5.24) into (5.21), and letting $\varepsilon \to 0$, we obtain (5.20).   ∎

Lemma 5.2 *If v_{z-w} is zero, then solutions of system (2.64)–(2.67) satisfy*

$$\int_{0}^{L}\int_{0}^{2\pi}\int_{0}^{1} v_z v_r \frac{\partial \tilde{V}_z}{\partial r} r \, dr \, d\theta \, dz \leq \frac{a}{2}\left(1+b\right)\int_{0}^{L}\int_{0}^{2\pi} v_{r-w}^{2} \, d\theta \, dz$$

$$+ \frac{a}{4}\left(1+\frac{1}{b}\right)\int_{0}^{L}\int_{0}^{2\pi}\int_{0}^{1}\left(\frac{\partial v_r}{\partial r}\right)^{2} r \, dr \, d\theta \, dz + \frac{1}{4a}\int_{0}^{L}\int_{0}^{2\pi}\int_{0}^{1}\left(\frac{\partial v_z}{\partial r}\right)^{2} r \, dr \, d\theta \, dz \tag{5.25}$$

for arbitrary positive constants a and b.

Proof. Since $\sup_\Omega \left| \frac{\partial \tilde{V}_z}{\partial r} \right| = 2$, we have

$$\int_0^L \int_0^{2\pi} \int_\varepsilon^1 v_z v_r \frac{\partial \tilde{V}_z}{\partial r} r dr d\theta dz \leq 2 \int_0^L \int_0^{2\pi} \int_\varepsilon^1 |v_z||v_r| r dr$$

$$\leq a \int_0^L \int_0^{2\pi} \int_\varepsilon^1 v_r^2 r dr d\theta dz + \frac{1}{a} \int_0^L \int_0^{2\pi} \int_\varepsilon^1 v_z^2 r dr d\theta dz. \quad (5.26)$$

We write

$$v_r(r, \theta, z, t) = v_{r-w} - \int_r^1 \frac{\partial v_r}{\partial r} dr$$

so that

$$v_r^2(r, \theta, z, t) = \left(v_{r-w} - \int_r^1 \frac{\partial v_r}{\partial r} dr \right)^2 \leq (1+b) v_{r-w}^2 + \left(1 + \frac{1}{b} \right) \left(\int_r^1 \frac{\partial v_r}{\partial r} dr \right)^2.$$

By the Schwartz inequality,

$$\left(\int_r^1 \frac{1}{\sqrt{r}} \sqrt{r} \frac{\partial v_r}{\partial r} dr \right)^2 \leq - \ln r \int_r^1 r \left(\frac{\partial v_r}{\partial r} \right)^2 dr$$

so we have that

$$r v_r^2(r, \theta, z, t) \leq (1+b) r v_{r-w}^2 - \left(1 + \frac{1}{b} \right) r \ln r \int_\varepsilon^1 r \left(\frac{\partial v_r}{\partial r} \right)^2 dr$$

where we have set $r = \varepsilon$ in the lower integral limit. We now get

$$\int_0^L \int_0^{2\pi} \int_\varepsilon^1 v_r^2 r dr d\theta dz \leq$$

$$\int_0^L \int_0^{2\pi} \int_\varepsilon^1 \left((1+b) r v_{r-w}^2 - \left(1 + \frac{1}{b} \right) r \ln r \int_\varepsilon^1 r \left(\frac{\partial v_r}{\partial r} \right)^2 dr \right) dr d\theta dz$$

$$= \frac{1}{2} (1 - \varepsilon^2)(1+b) \int_0^L \int_0^{2\pi} v_{r-w}^2 d\theta dz$$

$$+ \frac{1}{4} (1 - \varepsilon^2 + 2\varepsilon^2 \ln \varepsilon) \left(1 + \frac{1}{b} \right) \int_0^L \int_0^{2\pi} \int_\varepsilon^1 r \left(\frac{\partial v_r}{\partial r} \right)^2 dr d\theta dz. \quad (5.27)$$

For v_z, we have

$$v_z(r, \theta, z, t) = v_{z-w} - \int_r^1 \frac{\partial v_z}{\partial r} dr = - \int_r^1 \frac{\partial v_z}{\partial r} dr,$$

so we get

$$v_z^2(r, \theta, z, t) = \left(\int_r^1 \frac{1}{\sqrt{r}} \sqrt{r} \frac{\partial v_z}{\partial r} dr \right)^2 \le - \ln r \int_r^1 r \left(\frac{\partial v_z}{\partial r} \right)^2 dr.$$

and, finally

$$\int_0^L \int_0^{2\pi} \int_\varepsilon^1 v_z^2 r \, dr d\theta dz \le \int_0^L \int_0^{2\pi} \int_\varepsilon^1 -r \ln r \int_\varepsilon^1 r \left(\frac{\partial v_z}{\partial r} \right)^2 dr dr d\theta dz$$

$$= \frac{1}{4} \left(1 - \varepsilon^2 + 2\varepsilon^2 \ln \varepsilon \right) \int_0^L \int_0^{2\pi} \int_\varepsilon^1 r \left(\frac{\partial v_z}{\partial r} \right)^2 dr d\theta dz. \quad (5.28)$$

Inserting (5.27) and (5.28) into (5.26), and letting $\varepsilon \to 0$, yield (5.25).   ■

The conditions of Lemma 5.1 and 5.2 are assumed to hold throughout the analysis that follows, that is: $v_{\theta-w} = v_{z-w} = 0$, and; v_{r-w} satisfies (5.16).

5.3.4   Optimality

The following theorem incorporates the control design and optimality result.

Theorem 5.1 *The control*

$$v_{r-w} = -k\Delta p, \quad (5.29)$$

with $k \in \left(0, \frac{Re}{4}\right)$ and Re arbitrary, maximizes the cost functional

$$J(v_{r-w}) = \lim_{t \to \infty} \left[2\beta E(\mathbf{w}(t)) + \int_0^t h(\mathbf{w}(\tau)) d\tau \right] \quad (5.30)$$

where

$$\beta = \frac{2k}{\left(1 - \frac{4}{Re} k\right)}$$

and

$$h\left(\mathbf{w}\right) = \frac{2\beta}{Re}m\left(\mathbf{w}\right) + 2\beta \int_0^L \int_0^{2\pi} \int_0^1 v_z v_r \frac{\partial \tilde{V}_z}{\partial r} r \, dr \, d\theta \, dz$$

$$- \left(\frac{\beta}{2}\right)^2 \left(1 + \frac{2\beta}{Re}\right)^{-1} \int_0^L \int_0^{2\pi} \Delta p^2 \, d\theta \, dz - \int_0^L \int_0^{2\pi} v_{r-w}^2 \, d\theta \, dz. \quad (5.31)$$

Moreover, solutions of system (2.64)–(2.67) satisfy

$$h\left(\mathbf{w}\right) \le c_1 m\left(\mathbf{w}\right) - c_2 \int_0^L \int_0^{2\pi} \Delta p^2 \, d\theta \, dz - \frac{1}{2} \int_0^L \int_0^{2\pi} v_{r-w}^2 \, d\theta \, dz \qquad (5.32)$$

for arbitrary values of the control v_{r-w}, and with

$$c_1 = \frac{2\beta}{Re} + \max\left(\frac{1}{4}, 2\beta^2\right) > 0 \text{ and } c_2 = \left(\frac{\beta}{2}\right)^2 \left(1 + \frac{2\beta}{Re}\right)^{-1} > 0. \qquad (5.33)$$

Proof. By Lemma 5.1, we can write (5.31) as

$$h\left(\mathbf{w}\right) = -2\beta \dot{E}(\mathbf{w}) - \beta \int_0^L \int_0^{2\pi} v_{r-w} \Delta p \, d\theta \, dz - \frac{2\beta}{Re} \int_0^L \int_0^{2\pi} v_{r-w}^2 \, d\theta \, dz$$

$$- \left(\frac{\beta}{2}\right)^2 \left(1 + \frac{2\beta}{Re}\right)^{-1} \int_0^L \int_0^{2\pi} \Delta p^2 \, d\theta \, dz - \int_0^L \int_0^{2\pi} v_{r-w}^2 \, d\theta \, dz. \quad (5.34)$$

Inserting (5.34) into (5.30) we get

$$J\left(v_{r-w}\right) = \lim_{t \to \infty} \left[2\beta E\left(\mathbf{w}\left(t\right)\right) + \int_0^t \left(\left(-2\beta \dot{E}(\mathbf{w}\left(\tau\right))\right) \right. \right.$$

$$- \beta \int_0^L \int_0^{2\pi} v_{r-w} \Delta p \, d\theta \, dz - \frac{2\beta}{Re} \int_0^L \int_0^{2\pi} v_{r-w}^2 \, d\theta \, dz$$

$$\left. \left. - \left(\frac{\beta}{2}\right)^2 \left(1 + \frac{2\beta}{Re}\right)^{-1} \int_0^L \int_0^{2\pi} \Delta p^2 \, d\theta \, dz - \int_0^L \int_0^{2\pi} v_{r-w}^2 \, d\theta \, dz \right) d\tau \right]$$

$$= \lim_{t \to \infty} \left[2\beta E\left(\mathbf{w}\left(t\right)\right) - 2\beta \int_0^t \dot{E}(\mathbf{w}\left(\tau\right)) d\tau \right.$$

$$
-\left(1+\frac{2\beta}{Re}\right)\int_0^t\int_0^L\int_0^{2\pi}\left(v_{r-w}^2+\beta\left(1+\frac{2\beta}{Re}\right)^{-1}v_{r-w}\Delta p\right.
$$

$$
\left.+\left(\frac{\beta}{2}\right)^2\left(1+\frac{2\beta}{Re}\right)^{-2}\Delta p^2\right)d\theta dz d\tau\Bigg]
$$

$$
=2\beta E\left(\mathbf{w}\left(0\right)\right)
$$

$$
-\left(1+\frac{2\beta}{Re}\right)\lim_{t\to\infty}\int_0^t\int_0^L\int_0^{2\pi}\left(v_{r-w}+\frac{\beta}{2}\left(1+\frac{2\beta}{Re}\right)^{-1}\Delta p\right)^2 d\theta dz d\tau. \quad (5.35)
$$

The maximum of (5.30) is achieved when the integral in (5.35) is zero. Thus, (5.29) is the optimal control. Inequality (5.32) is obtained by applying Lemma 5.2 with $a=\frac{1}{4\beta}$ and $b=1$, to (5.31). $\blacksquare$

The objective of applying the control input (5.29) is to increase the value of $m(\mathbf{w})$. That this objective is targeted in the cost functional (5.30), is clear from inequality (5.32), which gives an upper bound on $h\left(\mathbf{w}\right)$ in terms of $m\left(\mathbf{w}\right)$. Thus, $h\left(\mathbf{w}\right)$ cannot be made large without making $m\left(\mathbf{w}\right)$ large, so the cost functional (5.30) is meaningful with respect to our objective. The cost functional also puts penalty on the output. Since the output is fed back to the control input, the output penalty works in conjunction with the input penalty to minimize control effort.

The next theorem writes the result of Theorem 5.1 on a form that puts emphasis on signal gains.

Theorem 5.2 *For all Re and $t\geq0$, solutions of system (2.64)–(2.67) satisfy*

$$
\lim_{t\to\infty}\underbrace{\frac{2\beta E\left(\mathbf{w}(t)\right)+\int_0^t g\left(\mathbf{w}\left(\tau\right)\right)d\tau}{2\beta E\left(\mathbf{w}\left(0\right)\right)+c_2\int_0^t\int_0^L\int_0^{2\pi}\Delta p^2 d\theta dz d\tau+\frac{1}{2}\int_0^t\int_0^L\int_0^{2\pi}v_{r-w}^2 d\theta dz d\tau}}_{\substack{\max\\ v_{r-w}\\ E(\mathbf{w}(0))\neq0}}=1, \quad (5.36)
$$

where

$$
g\left(\mathbf{w}\right)\leq c_1 m\left(\mathbf{w}\right). \quad (5.37)
$$

Furthermore, the maximum is achieved with the optimal control (5.29), for which solutions of the closed–loop system satisfy

$$
2\beta E\left(\mathbf{w}\left(t\right)\right)+c_1\int_0^t m\left(\mathbf{w}\left(\tau\right)\right)d\tau\geq
$$

$$2\beta E\left(\mathbf{w}\left(0\right)\right) + \left(\frac{3}{2} + \frac{2\beta}{Re}\right) \int_0^t \int_0^L \int_0^{2\pi} v_{r-w}^2 \, d\theta dz d\tau. \quad (5.38)$$

Proof. Consider the function

$$g\left(\mathbf{w}\right) \triangleq h\left(\mathbf{w}\right) + c_2 \int_0^L \int_0^{2\pi} \Delta p^2 \, d\theta dz + \frac{1}{2} \int_0^L \int_0^{2\pi} v_{r-w}^2 \, d\theta dz. \quad (5.39)$$

Integration of (5.39) with respect to time, and adding $2\beta E\left(\mathbf{w}\left(t\right)\right)$ to each side, gives

$$2\beta E\left(\mathbf{w}\left(t\right)\right) + \int_0^t g\left(\mathbf{w}\left(\tau\right)\right) d\tau = 2\beta E\left(\mathbf{w}\left(t\right)\right)$$

$$+ \int_0^t h\left(\mathbf{w}\left(\tau\right)\right) d\tau + c_2 \int_0^t \int_0^L \int_0^{2\pi} \Delta p^2 \, d\theta dz d\tau + \frac{1}{2} \int_0^t \int_0^L \int_0^{2\pi} v_{r-w}^2 \, d\theta dz d\tau. \quad (5.40)$$

The two first terms on the right hand side of (5.40) is $J\left(v_{r-w}\right)$ (without the limit), so inserting (5.35) we get

$$2\beta E\left(\mathbf{w}\left(t\right)\right) + \int_0^t g\left(\mathbf{w}\left(\tau\right)\right) d\tau$$

$$= 2\beta E\left(\mathbf{w}\left(0\right)\right) - \left(1 + \frac{2\beta}{Re}\right) \int_0^t \int_0^L \int_0^{2\pi} \left(v_{r-w} + \frac{\beta}{2}\left(1 + \frac{2\beta}{Re}\right)^{-1} \Delta p\right)^2 d\theta dz d\tau$$

$$+ c_2 \int_0^t \int_0^L \int_0^{2\pi} \Delta p^2 \, d\theta dz d\tau + \frac{1}{2} \int_0^t \int_0^L \int_0^{2\pi} v_{r-w}^2 \, d\theta dz d\tau. \quad (5.41)$$

Dividing both sides of (5.41) by

$$2\beta E\left(\mathbf{w}\left(0\right)\right) + c_2 \int_0^t \int_0^L \int_0^{2\pi} \Delta p^2 \, d\theta dz d\tau + \frac{1}{2} \int_0^t \int_0^L \int_0^{2\pi} v_{r-w}^2 \, d\theta dz d\tau,$$

assuming $E\left(\mathbf{w}\left(0\right)\right) \neq 0$, taking the limit as $t \to \infty$, and then taking the maximum value over v_{r-w}, we obtain

$$\underbrace{\lim_{t \to \infty} \frac{2\beta E\left(\mathbf{w}\left(t\right)\right) + \int_0^t g\left(\mathbf{w}\left(\tau\right)\right) d\tau}{2\beta E\left(\mathbf{w}\left(0\right)\right) + c_2 \int_0^t \int_0^L \int_0^{2\pi} \Delta p^2 \, d\theta dz d\tau + \frac{1}{2} \int_0^t \int_0^L \int_0^{2\pi} v_{r-w}^2 \, d\theta dz d\tau}}_{\substack{\max \\ v_{r-w} \\ E\left(\mathbf{w}\left(0\right)\right) \neq 0}}$$

$$
= 1 - \lim_{t \to \infty} \frac{\left(1 + \frac{2\beta}{Re}\right) \int_0^t \int_0^L \int_0^{2\pi} \left(v_{r-w} + \frac{\beta}{2}\left(1 + \frac{2\beta}{Re}\right)^{-1} \Delta p\right)^2 d\theta dz d\tau}{2\beta E\left(\mathbf{w}\left(0\right)\right) + c_2 \int_0^t \int_0^L \int_0^{2\pi} \Delta p^2 d\theta dz d\tau + \frac{1}{2} \int_0^t \int_0^L \int_0^{2\pi} v_{r-w}^2 d\theta dz d\tau} . \quad (5.42)
$$

$$
\underbrace{}_{\substack{\max \\ v_{r-w} \\ E\left(\mathbf{w}\left(0\right)\right) \neq 0}}
$$

Since the numerator of the last term in (5.42) is non-negative, and the denominator is strictly positive, the maximum on the right hand side of (5.42) is attained when the numerator is zero, which is for the optimal control (5.29). Thus, we obtain (5.36). Inequality (5.37) follows from (5.39) and (5.32). Inserting the optimal control into (5.41) by writing Δp in terms of v_{r-w} using (5.29), we obtain

$$
2\beta E\left(\mathbf{w}\left(t\right)\right) + \int_0^t g\left(\mathbf{w}\left(\tau\right)\right) d\tau = 2\beta E\left(\mathbf{w}\left(0\right)\right)
$$

$$
+ c_2 \left(\frac{2}{\beta}\right)^2 \left(1 + \frac{2\beta}{Re}\right)^2 \int_0^t \int_0^L \int_0^{2\pi} v_{r-w}^2 d\theta dz d\tau + \frac{1}{2} \int_0^t \int_0^L \int_0^{2\pi} v_{r-w}^2 d\theta dz d\tau. \quad (5.43)
$$

Inserting for c_2, as defined in (5.33), and using (5.37), we get (5.38). ∎

The result (5.36) was inspired by the work on optimal destabilization of linear systems reported in [106]. In view of (5.37), by maximizing the ratio in the curly brackets of (5.36), we make sure that the input and output signals are small compared to the internal states. This is equivalent to obtaining a large closed-loop gain. In addition, the theorem gives a lower bound on the states in terms of the control input for system (2.64)–(2.67) in closed loop with (5.29). Thus, it establishes the fact that the states cannot be small without the control input being small, and the control input cannot be made large without making the states large. As we shall see in our simulation study, this will lead to good mixing with low control effort.

5.3.5 Detectability of Mixing

Achieving optimality with static output feedback of Δp is remarkable. In this section we explain why this special output is strongly related to mixing and allows its enhancement. The next theorem establishes an open-loop property of system (2.64)–(2.67) that is reminiscent of an integral variant of input/output-to-state-stability (IOSS) for finite dimensional nonlinear systems.

Theorem 5.3 *If $Re \in (0,4)$, then solutions of system (2.64)–(2.67) satisfy*

$$c_3 \int_0^t m\left(\mathbf{w}\left(\tau\right)\right) d\tau \ \leq\ 2\beta E\left(\mathbf{w}\left(0\right)\right)$$

$$+\beta^2 \left(1+\frac{2\beta}{Re}\right)^{-1} \int_0^t \int_0^L \int_0^{2\pi} \Delta p^2 d\theta dz d\tau$$

$$+c_4 \int_0^t \int_0^L \int_0^{2\pi} v_{r-w}^2 d\theta dz d\tau, \qquad (5.44)$$

for all $t \geq 0$ and for arbitrary values of the control v_{r-w}, with

$$c_3 = \frac{\beta}{4}\left(\frac{4-Re}{Re}\right) > 0 \ \text{and} \ c_4 = 1+\beta\left(\frac{4+Re}{4-Re}\right) > 0.$$

Proof. From (5.30), (5.31) and (5.35), we get for all $t \geq 0$:

$$2\beta E\left(\mathbf{w}\left(t\right)\right) + \int_0^t \frac{2\beta}{Re} m\left(\mathbf{w}\left(\tau\right)\right) d\tau$$

$$\leq 2\beta E\left(\mathbf{w}\left(0\right)\right) - 2\beta \int_0^t \int_0^L \int_0^{2\pi} \int_0^1 v_z v_r \frac{\partial \tilde{V}_z}{\partial r} r dr d\theta dz d\tau$$

$$+ \left(\frac{\beta}{2}\right)^2 \left(1+\frac{2\beta}{Re}\right)^{-1} \int_0^t \int_0^L \int_0^{2\pi} \Delta p^2 d\theta dz d\tau + \int_0^t \int_0^L \int_0^{2\pi} v_{r-w}^2 d\theta dz d\tau. \quad (5.45)$$

Using Lemma 5.2 , we obtain

$$2\beta E\left(\mathbf{w}\left(t\right)\right) + \int_0^t \frac{2\beta}{Re} m\left(\mathbf{w}\left(\tau\right)\right) d\tau$$

$$\leq 2\beta E\left(\mathbf{w}\left(0\right)\right) + 2\beta \int_0^t \left(\frac{a}{2}\left(1+b\right) \int_0^L \int_0^{2\pi} v_{r-w}^2 d\theta dz\right.$$

$$+\frac{a}{4}\left(1+\frac{1}{b}\right) \int_0^L \int_0^{2\pi} \int_0^1 \left(\frac{\partial v_r}{\partial r}\right)^2 r dr d\theta dz + \frac{1}{4a} \int_0^L \int_0^{2\pi} \int_0^1 \left(\frac{\partial v_z}{\partial r}\right)^2 r dr d\theta dz\right) d\tau$$

$$+ \left(\frac{\beta}{2}\right)^2 \left(1+\frac{2\beta}{Re}\right)^{-1} \int_0^t \int_0^L \int_0^{2\pi} \Delta p^2 d\theta dz d\tau + \int_0^t \int_0^L \int_0^{2\pi} v_{r-w}^2 d\theta dz d\tau,$$

so it follows that

$$2\beta E\left(\mathbf{w}\left(t\right)\right) + \int\limits_0^t \frac{2\beta}{Re} m\left(\mathbf{w}\left(\tau\right)\right) d\tau$$

$$\leq 2\beta E\left(\mathbf{w}\left(0\right)\right) + \int\limits_0^t \frac{\beta}{2} \max\left(a\left(1+\frac{1}{b}\right),\frac{1}{a}\right) m\left(\mathbf{w}\left(\tau\right)\right) d\tau$$

$$+\left(\frac{\beta}{2}\right)^2 \left(1+\frac{2\beta}{Re}\right)^{-1} \int\limits_0^t \int\limits_0^L \int\limits_0^{2\pi} \Delta p^2 d\theta dz d\tau + \left(1+\beta a\left(1+b\right)\right) \int\limits_0^t \int\limits_0^L \int\limits_0^{2\pi} v_{r-w}^2 d\theta dz d\tau.$$

Rearranging the terms, we obtain

$$2\beta E\left(\mathbf{w}\left(t\right)\right) + \left(\frac{2\beta}{Re} - \frac{\beta}{2}\max\left(a\left(1+\frac{1}{b}\right),\frac{1}{a}\right)\right) \int\limits_0^t m\left(\mathbf{w}\left(\tau\right)\right) d\tau$$

$$\leq 2\beta E\left(\mathbf{w}\left(0\right)\right) + \left(\frac{\beta}{2}\right)^2 \left(1+\frac{2\beta}{Re}\right)^{-1} \int\limits_0^t \int\limits_0^L \int\limits_0^{2\pi} \Delta p^2 d\theta dz d\tau$$

$$+ \left(1+\beta a\left(1+b\right)\right) \int\limits_0^t \int\limits_0^L \int\limits_0^{2\pi} v_{r-w}^2 d\theta dz d\tau,$$

which is (5.44) for $a=1$ and $b=2\left(\frac{Re}{4-Re}\right)$. ∎

The significance of inequality (5.44) is that it provides a notion of detectability of internal states from the output Δp. In particular, if $m(\mathbf{w})$ is large, Δp must be large as well, or if Δp is small, so is $m(\mathbf{w})$. This is reminiscent of an integral variant of the IOSS property for finite-dimensional nonlinear systems, as presented in [89] (and motivated by earlier results in [122, 123]). In the case of (5.44) we have an integral-to-integral property (iiIOSS) with $m(\mathbf{w})$ as a measure of the states, so the "energy" of the states is bounded above by the "energy" of the input and output signals. With $E\left(\mathbf{w}\right)$ as a measure of the states, we can also find a uniform upper bound (as opposed to an "energy" upper bound) in terms of the input and output signals. That is, system (2.64)–(2.67) has the IOSS property, as stated formally in the next theorem.

Theorem 5.4 *For $Re \in (0,4)$, solutions of system (2.64)–(2.67) satisfy*

$$E\left(\mathbf{w}\left(t\right)\right)$$

$$\leq E\left(\mathbf{w}\left(0\right)\right) e^{-c_5 t} + \frac{1}{4c_5} \sup_{[0,t]}\left\{\int\limits_0^L \int\limits_0^{2\pi} \Delta p^2 d\theta dz\right\} + \frac{c_6}{c_5} \sup_{[0,t]}\left\{\int\limits_0^L \int\limits_0^{2\pi} v_{r-w}^2 d\theta dz\right\}$$

$$\tag{5.46}$$

for all $t \geq 0$ and for arbitrary values of the control v_{r-w}, with

$$c_5 = 2\max\left(\frac{4}{3Re} - 1, \frac{4 - Re}{4 + Re}\right) > 0 \text{ and } c_6 = \max\left(\frac{1}{4}, \frac{1}{4} + \frac{5Re - 4}{Re\,(4 - Re)}\right) > 0.$$

Proof. From (5.27), (5.28) and a similar derivation for v_θ, we have

$$2E\left(\mathbf{w}\right) \leq \frac{1}{2}\left(1 + b\right) \int_0^L \int_0^{2\pi} v_{r-w}^2 \, d\theta dz$$

$$+ \frac{1}{4} \int_0^L \int_0^{2\pi} \int_0^1 \left(\left(1 + \frac{1}{b}\right)\left(\frac{\partial v_r}{\partial r}\right)^2 + \left(\frac{\partial v_\theta}{\partial r}\right)^2 + \left(\frac{\partial v_z}{\partial r}\right)^2\right) r dr d\theta dz, \quad (5.47)$$

and therefore

$$2E\left(\mathbf{w}\right) \leq \frac{1}{2}\left(1 + b\right) \int_0^L \int_0^{2\pi} v_{r-w}^2 \, d\theta dz + \frac{1}{4}\left(1 + \frac{1}{b}\right) m\left(\mathbf{w}\right). \quad (5.48)$$

From Lemma 5.1 and (5.48) we get

$$\dot{E}(\mathbf{w}) \leq -2\left(\frac{4}{Re}\left(\frac{b}{1 + b}\right) - 1\right) E\left(\mathbf{w}\right)$$

$$- \frac{1}{2} \int_0^L \int_0^{2\pi} v_{r-w} \Delta p d\theta dz + \frac{2b - 1}{Re} \int_0^L \int_0^{2\pi} v_{r-w}^2 \, d\theta dz,$$

so that

$$\dot{E}(\mathbf{w}) \leq -2\left(\frac{4}{Re}\left(\frac{b}{1 + b}\right) - 1\right) E\left(\mathbf{w}\right)$$

$$+ \frac{1}{4} \int_0^L \int_0^{2\pi} \left(v_{r-w}^2 + \Delta p^2\right) d\theta dz + \frac{2b - 1}{Re} \int_0^L \int_0^{2\pi} v_{r-w}^2 \, d\theta dz.$$

Setting

$$b = \max\left(\frac{1}{2}, \frac{2Re}{4 - Re}\right)$$

we obtain

$$\dot{E}(\mathbf{w}) \leq -c_5 E\left(\mathbf{w}\right) + \frac{1}{4} \int_0^L \int_0^{2\pi} \Delta p^2 d\theta dz + c_6 \int_0^L \int_0^{2\pi} v_{r-w}^2 \, d\theta dz \quad (5.49)$$

with

$$c_5 = 2 \max \left(\frac{4}{3Re} - 1, \frac{4 - Re}{4 + Re} \right), \text{ and } c_6 = \max \left(\frac{1}{4}, \frac{1}{4} + \frac{5Re - 4}{Re\,(4 - Re)} \right).$$

Inequality (5.46) now follows from the comparison principle [90, Lemma C.5] (and the triangle inequality applied to the two last terms in (5.49)).   ∎

In Theorem 5.4, the notation $\sup_{[0,t]}$ denotes the essential supremum taken over the finite time interval $[0,t]$. The detectability properties stated in Theorems 5.3 and 5.4 indicate that our choice of sensing, Δp, is appropriate.

5.3.6   Numerical Simulations

The Computational scheme

The simulations are performed using a flow solver that is based on a second-order staggered grid discretization, second-order time advancement, and a Poisson equation for pressure, based on a scheme designed by Akselvoll and P. Moin [9]. The length of the cylinder is $L = 3\pi$ and the radius is $R = 1$. The grid is structured, single-block with cylindrical coordinates. It is uniform and periodic in z and θ with Fourier-modes 64 and 128 respectively, and linearly spaced with ratio 8 : 1 in the radial direction in order to achieve high resolution at the wall. The adaptive time step was in the range of 0.06-0.08 with constant CFL number 0.5 and constant 1 volume flux per unit span. The Reynolds number we used was $Re = 2100$ which is slightly higher than the limiting number $Re = 2000$ for nonlinear stability. We ran both the controlled and the uncontrolled case for about 110 time units starting from a statistically steady state flow field with control gain $k = 0.1$ in the controlled case. The initial flow field was obtained from a random perturbation of the parabolic profile over a large time interval using the uncontrolled case.

Measuring mixing

Figure 5.20 shows that our control results in an approximately 50% increase in the perturbation energy and 92% almost instantaneous increase in the enstrophy. While comparison based on perturbation energy is important as it is the part of the cost functional (5.30), enstrophy provides us with a measurement that is more closely related to mixing.

The instantaneous streamwise vorticity along a cross section of the pipe (Figure 5.21) also shows some promise for increased mixing with higher values of vorticity and more complex vortex structures in the controlled case than in the

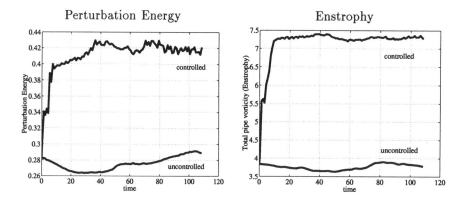

Figure 5.20: Perturbation energy and enstrophy [17].

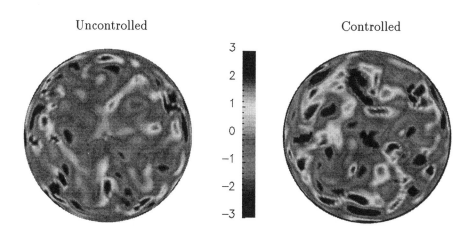

Figure 5.21: Streamwise vorticity [17].

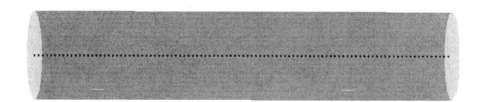

Figure 5.22: Initial particle distribution [17].

uncontrolled case. Vorticity is increased not only near the wall but everywhere in the pipe.

The method we use to quantify and visualize mixing is the tracking of dye in the flow. We consider the problem of mixing of a single fluid (or similar fluids) governed by the stretching and folding of material elements. We introduce passive tracer dye along the center of the pipe represented by a set of 100 particles, as shown in Figure 5.22. We trace the position of these particles using a particle-line method [88, 129]. The distance between neighboring particles is kept less than 0.1 by introducing new particles to halve the distance if necessary to obtain a connected dye surface at all time. As shown in Figure 5.23, the number of particles, that is, the length of the dye, increases in the controlled case at a much higher rate than in the uncontrolled case. Adding particles is not feasible computationally for an extended period of time. We stopped adding particles when their number reached two million ($t = 4$ in the controlled case and $t = 8$ in the uncontrolled case), but we continued tracing them. Figure 5.24 shows the distribution of particles inside the pipe. In the controlled case we obtain more uniform particle distribution even for smaller time.

Actuator distribution and bandwidth

Figure 5.25 shows the instantaneous pressure field in a cross section of the pipe along with the boundary velocity that is magnified 500 times for visualization. The control "blows in" when wall pressure is high and "sucks out" when wall pressure is low. Spatial changes in the control velocity are smooth and small, promising that a low number of actuators will suffice in practice. In order to investigate the density and bandwidth of sensors and actuators needed we calculate the power spectral densities of the control. The spectral plots alongside with the original signals are shown in Figure 5.26. Figures 5.26(a,b) show that only about 10-15 actuators/sensors are needed along the pipe length. Similarly, Figures 5.26(c,d) show that we need at most 15-20 actuators/sensors in the angular direction. That results in approximately 200 micro-actuators/sensors for the whole pipe surface. The time-frequency analysis in Figures 5.26(e,f) shows a bandwidth required for sensing/actuation of only 1.5Hz.

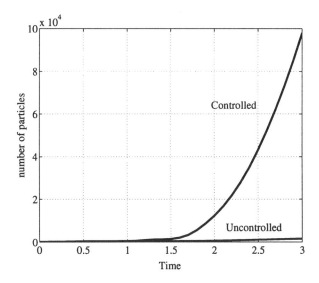

Figure 5.23: Length of dye as a function of time [17].

Uncontrolled, $t = 38$

Controlled, $t = 19$

Controlled, $t = 38$

Figure 5.24: Particle distribution [17].

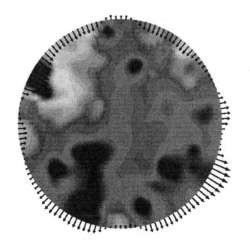

Figure 5.25: Instantaneous pressure field with controlled velocity (magnified) in a cross section of the pipe [17].

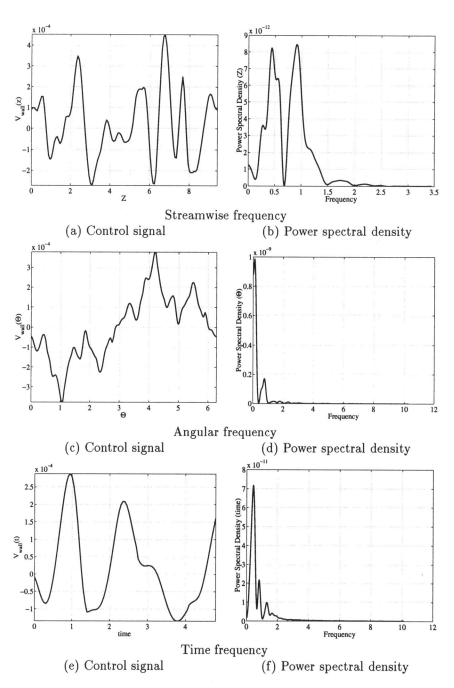

Figure 5.26: Spectral analysis of the control [17].

5.4   Particle Dispersion in Bluff Body Wakes

Motivated by the results in Sections 5.2 and 5.3, a simulation study was carried out in [6] investigating the feasibility of enhancing particle dispersion in the wake of a circular cylinder (2D) using similar, pressure-based, feedback control laws. The simulations were performed with FLUENT on the grid shown in Figure 4.27. For Reynolds numbers less than the critical Reynolds number $Re_c = 47$, the steady-state flow is symmetric about the streamwise axis. The objective is to initiate vortex shedding for a slightly subcritical case ($Re = 45$), causing increased particle dispersion in the wake.

The initial condition for the simulations is obtained by running FLUENT for 500 time units starting from a perturbed velocity field. In this case, which is subcritical, the disturbances are dampened out, as suggested by the time evolution of the lift coefficient, and confirmed by the high degree of symmetry in the vorticity map at $t = 500$, shown in Figure 5.27. The initial condition for the runs with feedback control is thus a *slightly* perturbed velocity field.

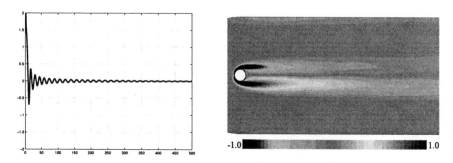

Figure 5.27: Lift coefficient for initial simulation (left graph), and vorticity field at the end of the initial simulation (right graph). The Reynolds number is $Re = 45$.

The feedback law has the form

$$\tau = k\left(a_1 \Delta p_1 + a_2 \Delta p_2 + a_3 \Delta p_3\right) \tag{5.50}$$

where k, a_1, a_2, and a_3 are constants, and $\Delta p_i = p_i^+ - p_i^-$, for $i = 1, 2, 3$ (see Figure 4.27 for the locations of the pressure sensors; actuation is applied through slot #2). Simulations are performed for three different values of the feedback gain k, as well as for the uncontrolled case.

The lift coefficient plots of Figure 5.28 show that in the uncontrolled case, the lift continues to decrease, while in the controlled cases, the lift coefficients increase and eventually reach a state at which the amplitude remains constant. Figure 5.29 shows the corresponding control signals, and Figure 5.30 shows vorticity

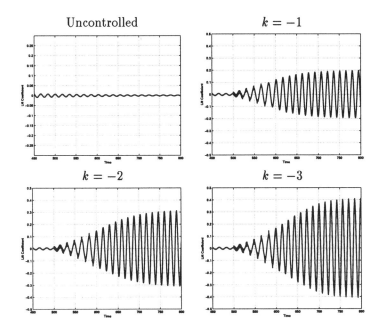

Figure 5.28: Lift coefficient for $Re = 45$.

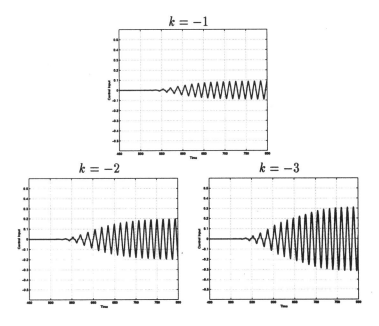

Figure 5.29: Control input for $Re = 45$.

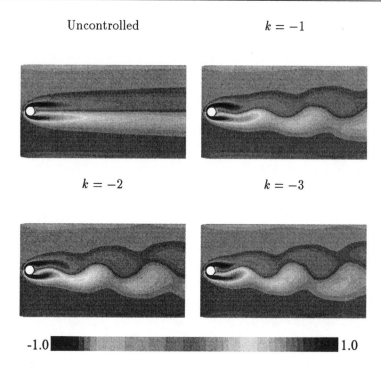

Figure 5.30: Vorticity field for $Re = 45$ at $t = 800$.

plots. The plots consistently show stronger vortices with increased feedback gain. In order to visualize particle dispersion, six strips of massless dye are put into the flow some distance upstream of the cylinder, and passively advected with the flow. A snapshot of the result is shown in Figure 5.31. It is clear that vortex shedding is initiated by our feedback control. Furthermore, the particles are dispersed more widely with increasing feedback gain.

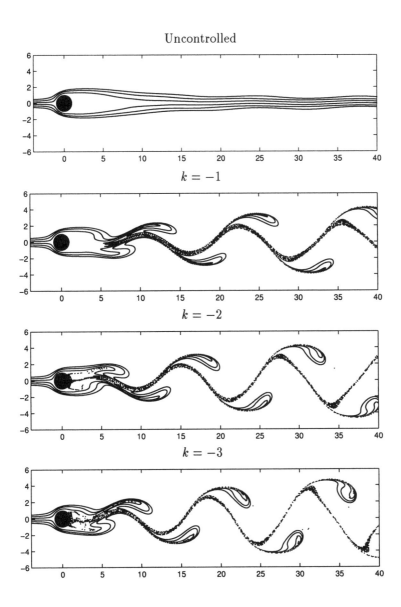

Figure 5.31: Particle distribution for $Re = 45$ at $t = 800$.

Chapter 6

Sensors and Actuators

6.1 Controlling Small-Scale Features

The theoretical and simulation results reviewed in this report are based on the ability to perform real-time distributed control. Distributed control is necessary due to the nature of drag increasing structures in the flow. In [34], high skin friction regions in turbulent flows are reported to occur near streamwise counter-rotating vortex pairs as illustrated in Figure 6.1. These vortex pairs bring high velocity fluid closer to the walls and thereby create local regions of high shear stress that significantly increase the total drag. In order to be able to control the appearance of such structures, we need to be able to sense and actuate at the same length scale, which decreases with increasing Reynolds number. The following example is presented in [34]. For an airflow at Reynolds number 10000, typical vortex pair streaks have width about 1 millimeter and length about 2 centimeters, appear at an approximate frequency of 100 Hertz, and has a life-time of about 1 millisecond. Thus, to have control authority over

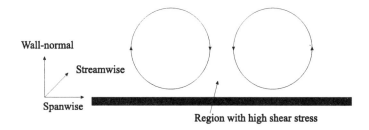

Figure 6.1: Counter-rotating vortex pair producing high shear stress regions [34].

Figure 6.2: Magnetic micromotor fabricated in nickel. Reproduced by courtesy of the University of Wisconsin, Madison.

these structures, the sensors and actuators must have sizes in the order of a few hundred micrometers. The technology to manufacture such small-scale sensors and actuators exists, and selected devices are reviewed in the next section.

6.2   Micro-Electro-Mechanical-Systems (MEMS)

6.2.1   General Properties of MEMS

The micromachining technology that was developed over the past decade or two, opens for fabrication of sensors and actuators on the micron scale. As an example of a micromachine, Figure 6.2 shows a photograph of a micromotor. Its rotor has a diameter of about the size of a human hair. This new technology is compatible with the integrated circuit (IC) technology, so that sensors and actuators can be integrated with controller logic into devices that can perform sensing, signal processing and actuation. Thus, for decentralized control strategies, the entire detection-decision-actuation process takes place locally, minimizing the need for wiring and data communication. By integrating a large number of such devices onto one chip, real-time distributed control can be realized. These systems are referred to as micro-electro-mechanical systems (MEMS). Typical micromachined devices are on the order of 100 microns, which is one or more orders of magnitude smaller than conventional sensors and actuators [66, 67]. The small size of the devices not only allows for high spatial resolution sensing and actuation, it also permits substantially faster actuation due to the increased frequency response that follows from decreased inertia. Although technology that permits distributed control of turbulent fluid flows exists, as documented through a number of successful prototype lab experiments [61, 132], commercial off-the-shelf systems are not available at this point. Nevertheless, the impressing progress in the field of MEMS manufacturing has been a driving force for modern flow control research. In the following sections, we will review some of the devices that are designed for flow control.

6.2.2 Micro Sensors

Pressure Sensors

Micro pressure sensors represent the most mature application of MEMS devices [67]. [99] describes the development and testing of an array of pressure sensors for obtaining the pressure distribution in a gaseous microchannel flow. Each individual sensor has a size of $250 \times 250 \mu m^2$. For the purpose of measuring intravascular blood pressure, even smaller devices have been manufactured [103], having a size of $100 \times 150 \mu m^2$.

Shear Stress Sensors

Shear stress sensors have been fabricated in sizes of $200 \times 200 \mu m^2$ [70, 100]. Such sensors have been integrated in large numbers on single chips to provide shear stress images of two-dimensional surfaces [75, 87]. They have also been integrated in flexible skins that can be glued on to curved surfaces [76], as shown in Figure 6.3.

6.2.3 Micro Actuators

Pumps

For applying wall transpiration by means of suction and blowing of fluid through tiny holes in the wall, micro pumps are needed. A number of micro pumps have been developed using a wide range of actuation principles (see [130, 135] for examples). Many of them consist of a cavity, with a diaphragm that seals the cavity, and inlets and outlets which are controlled by micro valves [120]. The size of these devices vary, but are at this point in the millimeter range.

Flaps

In [131], the use of micro flaps is suggested for pushing areas of high shear stress away from the wall in order to minimize overall drag. The size of this device, shown in Figure 6.4, is $300 \times 300 \mu m^2$. An experiment using this device, in conjunction with shear stress sensors, was conducted in [61], leading to a 2.5% reduction in skin friction.

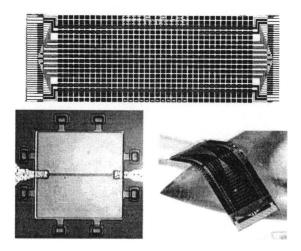

Figure 6.3: Upper picture: Flexible shear stress sensor array containing about 100 sensors. Its dimension is $1 \times 3cm^2$. Left picture: single shear stress sensor ($200 \times 200\mu m^2$). Right picture: The flexible skin resting on a conic object. The bending is caused by gravitation. All pictures are taken from [76] and shown by courtesy of Dr. Tai of Caltech, USA.

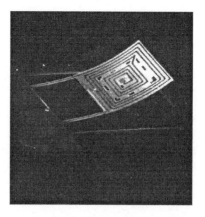

Figure 6.4: A magnetic flap for drag reduction in turbulent flows. Its dimension is $300 \times 300\mu m^2$. The picture is taken from [131] and shown by courtesy of Dr. Tsao of Caltech, USA.

Rotating Discs

An actuation method based on rotating discs was proposed in [84]. Due to the no-slip boundary condition, small rotating discs are capable of prescribing wall normal vorticity at the wall. Thus, in conjunction with wall normal transpiration, one obtains a two-component actuation device that completely specifies the near-wall flow field in the case of incompressible flow.

6.3   Concluding Remarks

The examples of micro sensors and actuators presented above suggest that distributed control at turbulent scales will be realizable in the near future. This will provide us with a means of testing the control laws reviewed in this report in the laboratory, and will pave the road towards developing commercial flow control systems.

Appendix A

Coefficients for the Ginzburg-Landau Equation

The numerical coefficients below are taken from [117, Appendix A], with modifications given by [96].

$$Re_c = 47 \tag{A.1}$$

$$x^t = 1.183 - 0.031i \tag{A.2}$$

$$\omega_0^t = 0.690 + 0.080i + (-0.00159 + 0.00447i)(Re - Re_c) \tag{A.3}$$

$$k_0^t = 1.452 - 0.844i + (0.00341 + 0.011i)(Re - Re_c) \tag{A.4}$$

$$\omega_{kk}^t = -0.292i \tag{A.5}$$

$$\omega_{xx}^t = 0.108 - 0.057i \tag{A.6}$$

$$k_x^t = 0.164 - 0.006i \tag{A.7}$$

$$\omega_0(x) = \omega_0^t + \frac{1}{2}\omega_{xx}^t \left(x - x^t\right)^2 \tag{A.8}$$

$$k_0(x) = k_0^t + k_x^t \left(x - x^t\right) \tag{A.9}$$

$$a_1(x) = -\omega_{kk}^t k_0(x) \tag{A.10}$$

$$a_2 = -\frac{1}{2}i\omega_{kk}^t \tag{A.11}$$

$$a_3 = -0.638 + 0.191i + (0.0132 - 0.00399i)(Re - Re_c) \tag{A.12}$$

$$a_4(x) = \left(\omega_0 + \frac{1}{2}\omega_{kk}^t k_0^2(x)\right)i \tag{A.13}$$

$$a_5 = -0.0225 + 0.0671i. \tag{A.14}$$

Based on these parameters, we obtain

$$a_{R_1}(x) = 0.242\,89 - 0.003212\,(Re - Re_c) + 1.752 \times 10^{-3}x \tag{A.15}$$

$$a_{I_1}(x) = 0.367\,39 + 0.00099572\,(Re - Re_c) + 4.788\,8 \times 10^{-2} x \quad \text{(A.16)}$$

$$a_{R_2} = -0.146 \quad \text{(A.17)}$$

$$a_{I_2} = 0 \quad \text{(A.18)}$$

$$a_{R_3} = -0.638 + 0.0132\,(Re - Re_c) \quad \text{(A.19)}$$

$$a_{I_3} = 0.191 - 0.00399\,(Re - Re_c) \quad \text{(A.20)}$$

$$
\begin{aligned}
a_{R_4}(x) = {} & 9.391\,7 \times 10^{-2} - 5.454\,1 \times 10^{-4}\,(Re - Re_c) \\
& -1.596\,8 \times 10^{-5}\,(Re - Re_c)^2 - 1.198\,5 \times 10^{-2} x \\
& +1.825\,7 \times 10^{-4}\,(Re - Re_c)\,x + 3.242\,2 \times 10^{-2} x^2 \quad \text{(A.21)}
\end{aligned}
$$

$$
\begin{aligned}
a_{I_4}(x) = {} & 0.457\,83 + 1.623\,0 \times 10^{-3}\,(Re - Re_c) \\
& +1.095\,3 \times 10^{-5}\,(Re - Re_c)^2 - 0.168\,04 x \\
& +5.207\,9 \times 10^{-4}\,(Re - Re_c)\,x + 5.371\,3 \times 10^{-2} x^2 \quad \text{(A.22)}
\end{aligned}
$$

$$a_{R_5} = -0.0255 \quad \text{(A.23)}$$

$$a_{I_5} = 0.0671. \quad \text{(A.24)}$$

Bibliography

[1] O.M. Aamo, M. Krstić, and T.R. Bewley, "Control of mixing by boundary feedback in 2d channel flow," submitted to *Automatica*.

[2] O.M. Aamo, M. Krstić, and T.R. Bewley, "Fluid mixing by feedback in Poiseuille flow," *Proceedings of the American Control Conference*, Arlington, Virginia, USA, 2001.

[3] O.M. Aamo, A. Balogh, and M. Krstić, "Optimal mixing by feedback in pipe flow," accepted for presentation at the *15th IFAC World Congress*, Barcelona, Spain, 2002.

[4] O.M. Aamo and M. Krstić, "Global stabilization of a nonlinear Ginzburg-Landau model of vortex shedding," submitted to the *European Journal of Control*.

[5] O.M. Aamo and M. Krstić, "Backstepping design for a semi-discretized Ginzburg-Landau model of vortex shedding," submitted to the *IEEE 2002 Conference on Decision and Control*.

[6] O.M. Aamo and M. Krstić, "Feedback control of particle dispersion in bluff body wakes," in preparation.

[7] F. Abergel and R. Temam, "On some control problems in fluid mechanics," *Theoret. and Comput. Fluid Dynamics*, vol. 1, pp. 303-325, 1990.

[8] R. Adams, *Sobolev Spaces*, Academic Press, New York, 1975.

[9] K. Akselvoll and P. Moin, "An efficient method for temporal integration of the Navier–Stokes equations in confined axisymmetric geometry," *Journal of Computational Physics*, vol. 125, pp. 454–463, 1996.

[10] B.D.O. Anderson and J.B. Moore, *Optimal Control: Linear Quadratic Methods*, Prentice-Hall, Inc. 1989.

[11] A.M. Annaswamy and A.F. Ghoniem, "Active control in combustion systems," *IEEE Control Systems*, vol. 15, no. 6, pp. 49-63, 1995.

[12] H. Aref, "Stirring by chaotic advection," *Journal of Fluid Mechanics*, vol. 143, pp. 1-21, 1984.

[13] H. Aref and G. Tryggvason, "Vortex dynamics of passive and active interfaces," *Physica D*, vol. 12, pp. 59-70, 1984.

[14] J. Baker, A. Armaou, and P.D. Christofides, "Drag reduction in incompressible channel flow using electromagnetic forcing," *Proceedings of the American Control Conference*, Chicago, Illinois, USA, 2000.

[15] A. Balogh, W-J. Liu, and M. Krstić, "Stability enhancement by boundary control in 2d channel flow–part I: Regularity of solutions," *Proceedings of the 38th Conference on Decision and Control*, Phoenix, Arizona, USA, 1999.

[16] A. Balogh, W-J. Liu, and M. Krstić, "Stability enhancement by boundary control in 2d channel flow," *IEEE Transactions on Automatic Control*, vol. 46, no. 11, pp. 1696–1711, 2001.

[17] A. Balogh, O.M. Aamo, and M. Krstić, "Optimal mixing enhancement in 3d pipe flow," submitted to *IEEE Transactions on Control Systems Technology*.

[18] A. Balogh, private communication.

[19] B. Bamieh, F. Paganini, and M. Dahleh, "Distributed control of spatially-invariant systems," submitted to *IEEE Transactions on Automatic Control*.

[20] B. Bamieh, I. Mezić, and M. Fardad, "A framework for destabilization of dynamical systems and mixing enhancement," *Proceedings of the 49th IEEE Conference on Decision and Control*, Orlando, Florida USA, 2001.

[21] B. Bamieh and M. Dahleh, "Energy amplification in channel flows with stochastic excitation," *Physics of Fluids*, vol. 13, no. 11, pp. 3258–3269, 2001.

[22] V. Barbu, "The time optimal control of Navier-Stokes equations," *Systems & Control Letters*, vol. 30, pp. 93-100, 1997.

[23] V. Barbu and S.S. Sritharan, "$H^{\infty}-$control theory of fluid dynamics," *R. Soc. Lond. Proc. Ser. A Math. Phys. Eng. Sci.*, vol. 454, no. 1979, pp. 3009-3033, 1998.

[24] T.R. Bewley and S. Liu, "Optimal and robust control and estimation of linear paths to transition," *Journal of Fluid Mechanics*, vol. 365, pp. 305-349, 1998.

[25] T.R. Bewley, "Flow Control: New Challenges for a New Renaissance," *Progress in Aerospace Sciences*, vol. 37, pp. 21-58, 2001.

[26] T.R. Bewley and P. Moin. Technical report TF-76, Stanford University, 1999.

[27] T.R. Bewley, P. Moin, and R. Temam, "DNS-based predictive control of turbulence: an optimal benchmark for feedback algorithms," *Journal of Fluid Mechanics*, vol. 447, pp. 179–225, 2001.

[28] T.R. Bewley, R. Temam, and M. Ziane, "A general framework for robust control in fluid mechanics," *Physica D*, vol. 138, pp. 360-392, 2000.

[29] T.R. Bewley and O.M. Aamo, "On the search for fundamental performance limitations in fluid-mechanical systems," *Proceedings of the 2002 ASME Fluids Engineering Division Summer Meeting, Montreal*, Quebec, Canada, 2002.

[30] T.R. Bewley, "Analysis of a versatile moving-belt mechanism for the control of wall-bounded flows," preprint.

[31] R.B. Bird, W.E. Stewart, and E.N. Lightfoot, *Transport Phenomena*, John Wiley & Sons, Inc., 1960.

[32] D.M. Bošković and M. Krstić, "Nonlinear stabilization of a thermal convection loop by state feedback," *Automatica*, vol. 37, no. 12, pp. 2033-40, 2001.

[33] K.M. Butler and B.F. Farrell, "Three-dimensional optimal perturbations in viscous shear flows," *Physics of Fluids A*, vol. 4, pp. 1637-1650, 1992.

[34] J. Cantwell, "Organized motion in turbulent flow," *Annual Review of Fluid Mechanics*, vol. 13, pp. 475-515, 1981.

[35] C. Canuto, M.Y. Hussaini, A. Quarteroni, and T.A. Zang, *Spectral Methods in Fluid Dynamics*. Springer-Verlag, 1988.

[36] M.H. Carpenter and C.A. Kennedy, "Fourth-order 2N Runge-Kutta schemes," *NASA technical memorandum*, no. 109112, 1994.

[37] W.-L. Chien, H. Rising, and J.M. Ottino, "Laminar mixing and chaotic mixing in several cavity flows," *Journal of Fluid Mechanics*, vol. 170, pp. 355-377, 1986.

[38] P.D. Christofides and A. Armaou, "Nonlinear control of Navier-Stokes equations," *Proceedings of the American Control Conference*, Philadelphia, Pennsylvania, 1998.

[39] P. Constantin and C. Foias, *Navier–Stokes Equations*, The University of Chicago Press, Chicago and London, 1988.

[40] J.-M. Coron, "On the controllability of the 2-D incompressible Navier-Stokes equations with the Navier slip boundary conditions," *ESAIM: Control, Optim. Cal. Var.*, vol. 1, pp. 35-75, 1996.

[41] J.-M. Coron, "On null asymptotic stabilization of the 2-D Euler equation of incompressible fluids on simply connected domains," *Prepublications 98-59 Université de Paris-Sud*, Mathematiques, 1998.

[42] L. Cortelezzi, K.H. Lee, J. Kim, and J.L. Speyer, "Skin-friction drag reduction via robust reduced-order linear feedback control," *Int. J. Comput. Fluid Dyn.*, vol. 8, no. 1-2, pp. 79-92, 1998.

[43] L. Cortelezzi and J.L. Speyer, "Robust reduced-order controller of laminar boundary layer transitions," *Physical Review E*, vol. 58, no. 2, 1998.

[44] D. D'Alessandro, M. Dahleh, and I. Mezić, "Control of fluid mixing using entropy methods," *Proceedings of the American Control Conference*, Philadelphia, Pennsylvania, 1998.

[45] D. D'Alessandro, M. Dahleh, and I. Mezić, "Control of mixing in fluid flow: A maximum entropy approach," *IEEE Transactions on Automatic Control*, vol. 44, no. 10, pp.1852-1863, 1999.

[46] R. Dautray and J.L. Lions, *Mathematical analysis and numerical methods for science and technology, Vol.2, Functional and variational methods*, Springer–Verlag, Berlin, 1992.

[47] H.O. Fattorini and S.S. Sritharan, "Optimal control problems with state constraints in fluid mechanics and combustion," *Appl. Math. Optim.*, vol. 38, no. 2, pp. 159-192, 1998.

[48] E. Fernández-Cara and M. González-Burgos, "A result concerning controllability for the Navier-Stokes equations," *SIAM J. Control Optim.*, vol. 33, no. 4, pp. 1061-1070, 1995.

[49] J.H. Ferziger and M. Perić, *Computational Methods for Fluid Dynamics*, Springer-Verlag, 1999.

[50] A.V. Fursikov, M.D. Gunzburger, and L.S. Hou, "Boundary value problems and optimal boundary control for the Navier-Stokes system: the two-dimensional case," *SIAM J. Control Optim.*, vol. 36, pp. 852-894, 1998.

[51] J.G. Franjione and J.M. Ottino, "Feasibility of numerical tracking of material lines and surfaces in chaotic flows," *Physics of Fluids*, vol. 30, pp. 3641-3643, 1987.

[52] G.F. Franklin, J.D. Powell, and A. Emami-Naeini, *Feedback Control of Dynamic Systems, 3rd edition*, Addison-Wesley, Inc., 1994.

[53] M. Gad-el-Hak, "Modern developments in flow control," *Appl. Mech. Rev.*, vol. 49, no. 7, pp. 365-379, 1996.

[54] M. Gad-el-Hak, *Flow Control: Passive, Active and Reactive Flow Management*, Cambridge University Press, 2000.

[55] A.F. Ghoniem and K.K. Ng, "Numerical study of the dynamics of a forced shear layer," *Physics of Fluids*, vol. 30, no. 3, pp. 706-721, 1987.

[56] M. Green and D.J.N. Limebeer, *Linear robust control*, Prentice-Hall, Inc., 1995.

[57] J. Guckenheimer and P. Holmes, *Nonlinear Oscillations, Dynamical Systems, and Bifurcations of Vector Fields*, Springer-Verlag, 1983.

[58] M.D. Gunzburger, L. Hou, and T.P. Svobodny, "A numerical method for drag minimization via the suction and injection of mass through the boundary," In *Stabilization of Flexible Structures* (ed. J.P. Zolesio), Springer-Verlag, 1990.

[59] M.D. Gunzburger (Editor), *Flow Control*, Springer-Verlag, 1995.

[60] M.D. Gunzburger and H.C. Lee, "Feedback control of Karman vortex shedding," *Transactions of the ASME*, vol. 63, pp. 828–835, 1996.

[61] B. Gupta, R. Goodman, F. Jiang, Y-C. Tai, S. Tung, and C-M. Ho, "Analog VLSI system for active drag reduction," *IEEE Micro*, October, 1996.

[62] G. Haller and A.C. Poje, "Finite time transport in aperiodic flows," *Physica D*, vol. 119, pp.352-380, 1998.

[63] G. Haller, "Finding finite-time invariant manifolds in two-dimensional velocity fields," *Chaos*, vol.10, no. 1, 2000.

[64] G. Haller and G. Yuan, "Lagrangian coherent structures and mixing in two-dimensional turbulence," *Physica D*, vol. 147, pp. 352-370, 2000.

[65] G. Haller, "Distinguished material surfaces and coherent structures in three-dimensional fluid flows," *Physica D*, vol. 149, pp. 248-277, 2001.

[66] C-M. Ho and Y-C. Tai, "REVIEW: MEMS and its applications for flow control," *Journal of Fluids Engineering*, vol. 118, pp. 437-447, 1996.

[67] C-M. Ho and Y-C. Tai, "Micro-electro-mechanical systems (MEMS) and fluid flows," *Annu. Rev. Fluid Mech.*, vol. 30, pp. 579-612, 1998.

[68] P. Holmes, J.L. Lumley, and G. Berkooz, *Turbulence, coherent structures, dynamical systems and symmetry*, Cambridge University Press, 1996.

[69] L.S. Hou and Y. Yan, "Dynamics for controlled Navier-Stokes systems with distributed controls," *SIAM J. Control Optim.*, vol. 35, no. 2, pp. 654-677, 1997.

[70] J-B. Huang, C-M. Ho, S. Tung, C. Liu, and Y-C. Tai, "Micro thermal shear stress sensor with and without cavity underneath," *Proceedings of the 8th Annual International Workshop on MEMS*, Amsterdam, 1995.

[71] P. Huerre and P.A. Monkewitz, "Local and global instabilities in spatially developing flows," *Annu. Rev. Fluid Mech.*, vol. 22, pp. 473–537, 1990.

[72] M. Högberg and T.R. Bewley, "Spatially-compact convolution kernels for decentralized control and estimation of transition in plane channel flow," preprint.

[73] O.Y. Imanuvilov, "On exact controllability for the Navier-Stokes equations," *ESAIM: Control, Optim. Cal. Var.*, vol. 3, pp. 97-131, 1998.

[74] K. Ito and S. Kang, "A dissipative feedback control synthesis for systems arising in fluid dynamics," *SIAM J. Control Optim.*, vol. 32, no. 3, pp. 831-854, 1994.

[75] F. Jiang, Y-C. Tai, B. Gupta, R. Goodman, S. Tung, J-B. Huang, and C-M. Ho, "A surface-micromachined shear stress imager," *Proceedings of the 9th Annual International Workshop on MEMS*, San Diego, USA, 1996.

[76] F. Jiang, Y-C. Tai, K. Walsh, T. Tsao, G-B. Lee, and C-M. Ho, "A flexible MEMS technology and its first application to shear stress sensor skin," *Proceedings of the 10th Annual International Workshop on MEMS*, Nagoya, Japan, 1997.

[77] J. Jiménez, "Transition to turbulence in two-dimensional Poiseuille flow," *Journal of Fluid Mechanics*, vol. 218, pp. 265-297, 1990.

[78] S.S. Joshi, J.L. Speyer, and J. Kim, "A systems theory approach to the feedback stabilization of infinitesimal and finite-amplitude disturbances in plane Poiseuille flow," *Journal of Fluid Mechanics*, vol. 332, pp. 157-184, 1997.

[79] S.S. Joshi, J.L. Speyer, and J. Kim, "Finite dimensional optimal control of Poiseuille flow," *Journal of Guidance, Control, and Dynamics*, vol. 22, no. 2, pp. 340-348, 1999.

[80] M. Jovanović and B. Bamieh, "The spatio-temporal impulse response of the linearized Navier-Stokes equations," *Proceedings of the 2001 American Control Conference*, Arlington, Virginia, USA, 2001.

[81] M. Jovanović and B. Bamieh, "Modeling flow statistics using the linearized Navier-Stokes equations," *Proceedings of the 40th IEEE Conference on Decision and Control*, Orlando, Florida, USA, 2001.

[82] V. Jurdjevic and J.P. Quinn, "Controllability and stability," *Journal of Differential Equations*, vol. 28, pp. 381-389, 1978.

[83] S.M. Kang, V. Ryder, L. Cortelezzi, and J.L. Speyer, "State-space formulation and controller design for three-dimensional channel flows," *Proceedings of the American Control Conference*, San Diego, California, 1999.

[84] L. Keefe, "A normal vorticity actuator for near-wall modification of turbulent shear flows," *AIAA*, 1997.

[85] D.V. Khakhar, H. Rising, and J.M. Ottino, "Analysis of chaotic mixing in two model systems," *Journal of Fluid Mechanics*, vol. 172, pp. 419-451, 1986.

[86] H.K. Khalil, *Nonlinear Systems, second edition*, Prentice-Hall, Inc. 1996.

[87] M. Kimura, S. Tung, J. Lew, and C-M. Ho, "Shear stress imaging micro chip for detection of high shear stress regions," *Proceedings of the 3rd ASME/JSME Joint Fluids Engineering Conference*, San Francisco, USA, 1999.

[88] T.S. Krasnopolskaya, V.V. Meleshko, G.W.M. Peters, and H.E.H Meijer, "Mixing in Stokes flow in an annular wedge cavity," *Eur. J. Mech. B/Fluids*, 18 (1999), 793-822.

[89] M. Krichman, E.D. Sontag, and Y. Wang, "Input-output-to-state stability," *SIAM Journal on Control and Optimization*, vol. 39, no. 6, pp. 1874–1928, 2001.

[90] M. Krstić, I. Kanellakopoulos, P. V. Kokotović, *Nonlinear and Adaptive Control Design*, John Wiley & Sons, Inc., 1995.

[91] M. Krstić, "On global stabilization of Burgers' equation by boundary control," *Systems & Control Letters*, vol. 37, pp. 123-141, 1999.

[92] O.A. Ladyzhenskaya, *The Mathematical Theory of Viscous Incompressible Flow*, Second English edition, Gordon and Breach, Science Publisher, Inc., New York, 1969.

[93] O.A. Ladyzhenskaya, V.A. Solonnikov, N.N. Ural'ceva, *Linear and Quasilinear Equations of Parabolic Type*, Translations of AMS, vol. 23, 1968.

[94] J.E. Lagnese, D.L. Russell, and L. White (Editors), *Control and optimal design of distributed parameter systems*, Springer-Verlag, 1995.

[95] E. Lauga and T.R. Bewley, "H_∞ control of linear global instability in models of non-parallel wakes," *Proceedings of the Second International Symposium on Turbulence and Shear Flow Phenomena*, Stockholm, Sweden, 2001.

[96] E. Lauga, private communication, 2002.

[97] C.W. Leong and J.M. Ottino, "Experiments on mixing due to chaotic advection in a cavity," *Journal of Fluid Mechanics*, vol. 209, pp. 463-499, 1989.

[98] J.L. Lions and E. Magenes, *Non-homogeneous Boundary value Problems and Applications, Vol.I*, Springer–Verlag, Berlin, Heidelberg, New York, 1972.

[99] J. Liu, Y-C. Tai, and C-M. Ho, "MEMS for pressure distribution studies of gaseous flows in microchannels," *Proceedings of the 8th Annual International Workshop on MEMS*, Amsterdam, 1995.

[100] C. Liu, J-B. Huang, Z. Zhu, F. Jiang, S. Tung, Y-C. Tai, and C-M. Ho, "A micromachined flow shear-stress sensor based on thermal transfer principles," *Journal of Microelectromechanical systems*, vol. 8, no. 1, 1999.

[101] W.J. Liu and M. Krstić, "Stability enhancement by boundary control in the Kuramoto–Sivashinsky equation," *Nonlinear Analysis*, vol. 43, pp. 485–583, 2000.

[102] N. Malhotra, I. Mezić, and S. Wiggins, "Patchiness: A new diagnostic for Lagrangian trajectory analysis in time-dependent fluid flows," *International Journal of Bifurcation and Chaos*, vol. 8, no. 6, pp. 1053-1093, 1998.

[103] P. Melvås, E. Kälvesten, and G. Stemme, "A surface micromachined resonant beam pressure sensor," *Proceedings of the 14th Annual Conference on MEMS*, Interlaken, Switzerland, 2001.

[104] I. Mezić, *On geometrical and statistical properties of dynamical systems: Theory and applications*, Ph.D. thesis, California Institute of Technology, 1994.

[105] I. Mezić and S. Wiggins, "A method for visualization of invariant sets of dynamical systems based on the ergodic partition," *Chaos*, vol. 9, no. 1, pp.213-218, 1999.

[106] I. Mezić, "Nonlinear dynamics and ergodic theory methods in control of fluid flows: theory and applications," *Proceedings of the 2001 AFOSR Workshop on Dynamics and Control*, 2001.

[107] P.D. Miller, C.K.R.T. Jones, A.M. Rogerson, and L.J. Pratt, "Quantifying transport in numerically generated velocity fields," *Physica D*, vol. 110, pp. 105-122, 1997.

[108] B.R. Noack, I. Mezić, and A. Banaszuk, "Controlling vortex motion and chaotic advection," *Proceedings of the 39th IEEE Conference on Decision and Control*, Sydney, Australia, December 11-15, 2000.

[109] S.A. Orzag, "Accurate solution of the Orr-Sommerfeld stability equation," *Journal of Fluid Mechanics*, vol. 50, part 4, pp. 689-703, 1971.

[110] J.M. Ottino, *The kinematics of mixing: stretching, chaos, and transport*, Cambridge University Press, 1989.

[111] J.M. Ottino, "Mixing, chaotic advection, and turbulence," *Annu. Rev. Fluid Mech.*, vol. 22, pp. 207-53, 1990.

[112] R.L. Panton, *Incompressible flow, Second Edition*, John Wiley & Sons, Inc., 1996.

[113] D.S. Park, D.M. Ladd, and E.W. Hendricks, "Feedback control of von Kármán vortex shedding behind a circular cylinder at low Reynolds numbers," *Physics of Fluids*, vol. 6, no. 7, pp. 2390–2405, 1994.

[114] A.C. Poje, G. Haller, and I. Mezić, "The geometry and statistics of mixing in aperiodic flows," *Physics of Fluids*, vol. 11, no. 10, 1999.

[115] A.C. Poje and G. Haller, "Geometry of cross-stream mixing in a double-gyre ocean model," *Journal of Physical Oceanography*, vol. 29, pp. 1469-1665, 1999.

[116] V. Rom-Kedar, A. Leonard, and S. Wiggins, "An analytical study of transport, mixing and chaos in an unsteady vortical flow," *Journal of Fluid Mechanics*, vol. 214, pp. 347-394, 1990.

[117] K. Roussopoulos and P. A. Monkewitz, "Nonlinear modelling of vortex shedding control in cylinder wakes," *Physica D*, 97, pp. 264–273, 1996.

[118] B.L. Rozhdestvensky and I.N. Simakin, "Secondary flows in a plane channel: their relationship and comparison with turbulent flows," *Journal of Fluid Mechanics*, vol. 147, pp. 261-289, 1984.

[119] W. Rudin, *Functional Analysis*, McGraw–Hill Book Company, New York, 1973.

[120] M.T.A. Saif, B.E. Alaca, and H. Sehitoglu, "Analytical modeling of electrostatic membrane actuator for micro pumps," *IEEE Journal of Microelectromechanical Systems*, vol. 8, no. 3, pp. 335-345, 1999.

[121] S. Skogestad and I. Postlethwaite, *Multivariable Feedback Control, Analysis and Design*, John Wiley & Sons, Inc., 1996.

[122] E.D. Sontag and Y. Wang, "Output-to-state stability and detectability of nonlinear systems," *Systems & Control Letters*, vol. 29, pp. 279–290, 1997.

[123] E.D. Sontag, "Comments on integral variants of ISS," *Systems & Control Letters*, vol. 34, pp. 93–100, 1998.

[124] S.S. Sritharan, "Dynamic programming of the Navier-Stokes equations," *Systems & Control Letters*, vol. 16, pp. 299-307, 1991.

[125] S.S. Sritharan, "Optimal control of viscous flows." *SIAM*, 1998.

[126] P.D. Swanson and J.M. Ottino, "A comparative computational and experimental study of chaotic mixing of viscous fluids," *Journal of Fluid Mechanics*, vol. 213, pp. 227-249, 1990.

[127] R. Temam, *Navier–Stokes Equations: Theory and Numerical Analysis*, Third (Revised) edition, North–Holland Publishing Company, Amsterdam, 1984.

[128] R. Temam, *Navier–Stokes Equations and Nonlinear Functional Analysis*, Second edition, SIAM, Philadelphia, 1995.

[129] A.A. Ten, Y.Y. Podladchikov, D.A. Yuen, T.B. Larsen, and A.V. Malevsky, "Comparison of mixing properties in convection with the particle-line method," *Geophys. Res. Lett.*, vol. 25, pp. 3205-3208, 1998.

[130] J-H. Tsai and L. Lin, "A thermal bubble actuated micro nozzle-diffuser pump," *Proceedings of the 14th Annual Conference on MEMS*, Interlaken, Switzerland, 2001.

[131] T. Tsao, C. Liu, Y.C. Tai, and C.M. Ho, "Micromachined magnetic actuators for active fluid control," *ASME Application of Microfabrication to Fluid Mechanics*, Chicago, November 6-11, 1994.

[132] T. Tsao, F. Jiang, R. Miller, Y-C. Tai, B. Gupta, R. Goodman, S. Tung, and C-M. Ho, "An integrated MEMS system for turbulent boundary layer control," *Proceedings of the 1997 International Conference on Solid-State Sensors and Actuators*, Chicago, June 16-19, 1997.

[133] S. Wiggins, *Introduction to Applied Nonlinear Dynamical Systems and Chaos*, Springer-Verlag, 1990.

[134] S. Wiggins, *Chaotic Transport in Dynamical Systems*, Springer-Verlag, 1992.

[135] K-S. Yun, I-J. Cho, J-U. Bu, G-H. Kim, Y-S. Jeon, C-J. Kim, and E. Yoon, "A micropump driven by continuous electorwetting actuation for low voltage and low power operations," *Proceedings of the 14th Annual Conference on MEMS*, Interlaken, Switzerland, 2001.

[136] K. Zhou, J.C. Doyle, and K. Glover, *Robust Optimal Control*, Prentice-Hall, Inc., 1995.

Index